Hans Wolfgang Kölmel

Atlas of Cerebrospinal Fluid Cells

Second, Enlarged Edition

With 251 Figures, 139 in Color

Springer-Verlag
Berlin Heidelberg New York 1977

Dr. Hans Wolfgang Kölmel
Freie Universität Berlin, Klinikum Westend
Neurologische Abteilung
Spandauer Damm 130, D-1000 Berlin 19

ISBN-13: 978-3-642-66625-4 e-ISBN-13: 978-3-642-66623-0
DOI: 10.1007/978-3-642-66623-0

Library of Congress Cataloging in Publication Data. Kölmel, Hans Wolfgang, 1944–. Atlas of cerebrospinal fluid cells. Bibliography: p. Includes index. 1. Cerebrospinal fluid–Examination. 2. Diagnosis, Cytologic–Atlases. 3. Cerebrospinal fluid–Diseases–Atlases. I. Title. [DNLM: 1. Cerebrospinal fluid–Cytology–Atlases. 2. Cytodiagnosis–Atlases. QY17 K77a] RB55.K63 1977 616.8'04'7582 77-3578

Softcover reprint of the hardcover 2nd edition 1977

Würzburg
2121/3140-543210

Preface

The demand for an atlas of CSF cytology has made a second edition necessary very soon. The number of those who have learned to appreciate the value of this simple diagnostic procedure has apparently increased. Optimal methods of cell preparation, limited technical requirements and immediate availability of the results have all contributed to this development.

In the current edition, primary emphasis has again been placed on selection of informative illustrations and on quality of reproduction. Some figures have been replaced, and some added. The text has been expanded and improved where it seemed appropriate. Malignant lymphomas are treated as a separate topic, rather than as part of the chapter on hemoblastoses. I should like to thank Prof. Dr. Hansjörg Riehm of the pediatric department at Klinikum Charlottenburg, Freie Universität Berlin (Director: Prof. Dr. Hans Helge), for his suggestion in this regard, as well as for the material for figures on plate 47 (a–d).

My technical assistent, Mrs. Christine Ligner, again deserves my special thanks. Her suggestions and her careful cytologic work have been significant contributions to the success of this atlas.

Berlin, February, 1977 Hans W. Kölmel

Foreword to the First Edition

In taking advantage of the abundant laboratory material of the neurological clinic of Heidelberg University and of the neurology department of Klinikum Charlottenburg of the Freie Universität Berlin, Dr. KÖLMEL has succeeded in producing a clinically representative cross-section of the cytologic phenoma in cerebrospinal fluid. There has long been a need for such an atlas.
An introductory presentation on cell collection techniques and staining methods is both brief and precise. The excellent illustrations, which make up the most extensive part of the work, are the results of a sedimentation chamber technique. Dr. KÖLMEL has developed a simple method to the point that it allows optimal reproduceability. If cytodiagnosis of the CSF—the most difficult and problem-ridden area of exfoliative cytology—achieves greater attention and wider application in the future, it will certainly be to the credit of the author and his meticulous efforts.
I am convinced that the demand for this volume will be great.

JOHANNES SAYK
Director of the Department of Neurology
The University of Rostock

Preface to the First Edition

Cytologic examination of the cerebrospinal fluid is a technically simple but productive diagnostic procedure. This book attempts to provide guidance to neophytes in the field of cytology and hopes to provide stimulation to the more experienced.
In microscopy, as in other scientific fields, photography is superior to even the best description. However, illustrations are always restricted to showing one aspect of the great variety of cytologic phenomena. Knowledge of the range of possibilities is a matter of experience, but it is beyond the scope of this atlas to provide such experience. First, I wish to thank Prof. Dr. JOHANNES SAYK, director of the Department of Neurology of Rostock University, for his inspiration and advise, as well as for the introduction to this volume. In the same way, I must thank Prof. Dr. DIETER JANZ, head of the neurology department of Klinikum Charlottenburg, Freie Universität Berlin, who liberally supported my work in his department. I also wish to thank my colleagues at the university clinics in Heidelberg and Berlin, who aided me by placing case histories at may disposal. Prof. Dr. HEINZ GÄNSHIRT, director of the neurology department of Heidelberg University, and Prof. Dr. Dr. ROLF WÜLLENWEBER, head of the department of neurosurgery, Klinikum Charlottenburg, Freie Universität Berlin, deserve special thanks in this regard. Prof. Dr. BRUNO CHONÉ of the radiology department at Heidelberg University, provided assistance with cytologic techniques, especially in the fields of autoradiography and fluorescent-stain microscopy. Dr. GERT POTZ of the neurology department of Klinikum Mannheim, Heidelberg University (Prof. Dr. OTTO HALLEN), also cooperated in many efforts and provided the slide preparations for figures on plate 30 (a–d), 35 (c, d), and 42 (a, b). Mr. WALTHER FÖRSTER of Leitz Inc., Wetzlar, and Prof. ERNST WEBER of the Hochschule für Bildende Künste, Berlin, aided me in questions concerning microscopy and photography. Mr. FRANK C. DOUGHERTY translated the German text into English. Finally, I wish to express particular thanks to my technical assistent, Mrs. CHRISTINE LIGNER, who worked faithfully with me on the innumerable tasks which this atlas required.

Berlin, February, 1976 HANS W. KÖLMEL

Contents

1. Cytologic Techniques

1.1. Cerebrospinal Fluid Cell Preparation

The study of the cerebrospinal fluid (CSF) has only recently come into its own in the extensive field of exfoliative cytology. Although cytologic examination became a routine procedure shortly after the introduction of the technically simple lumbar puncture by QUINCKE [154] in 1891 [9, 13, 58, 65, 80, 157, 222, 231], CSF cytology has only in the last 20 years reached a point where cell examination can be of considerable clinical value [19, 22, 95, 121, 138, 162, 177, 223]. This recent advance is mainly due to the development of new slide preparation methods better suited to the special properties of the CSF. The following points must be considered in the interest of diagnostically useful preparation:

1. Only a limited amount of CSF—usually not more than 2–3 cm^3—is available for each cytologic examination.
2. A smear of a fresh, untreated CSF sample is of limited quality in comparison with specimens of other material. The cell content of a single specimen is usually too low, and the smear alone is capable of producing mechanical alterations in the cells. Both factors significantly reduce the probability of a reliable diagnosis.
3. The cells of the CSF are also extremely sensitive to thermal and chemical influences. Because of its low protein content, the CSF offers an unsuitable medium for living cells. Material exfoliated from the epithelium of the ventricles, from the leptomeninges, or from tumors endure adverse conditions before arriving at the lumbar area, the usual site of puncture. Low oncotic pressure causes some cells to swell, some to perish and others to become unrecognizable.

Rapid preservation is an important consideration for any cell preparation, not only for that of the CSF. Slide preparation must be quick and must attain a high concentration of undamaged cells from a very limited quantity of fluid. The development of techniques fulfilling these requirements has been the object of considerable experimental effort. For decades, the standard procedure involved centrifugation of the specimen followed by a smear of the sediment. However, this method is not capable of meeting the standards of modern cytology. Even a reduction of the rotation rate of the centrifuge to less than 1,000 rpm is not sufficient to prevent obvious morphologic alteration of the cells. Optimal cytodiagnosis under these conditions is impossible. However, this method of preparation is the least complicated, and it is still practiced in many laboratories. In most cases in which electron microscopy is necessary, centrifugation will remain the method of choice. Damage to cells can be reduced by adding preservatives (osmium oxide, for example) to the specimen immediately after collection [73, 77, 80, 172, 218].

Inspired by earlier experiments [5]. SCHÖNENBERG [176, 177] placed a slide diagonally in a test tube filled with CSF and allowed spontaneous sedimentation of the cells. He thereby eliminated the accelerative forces of the centrifuge but had to accept sedimentation periods of several hours. SAYK [162] accelerated the process of spontaneous sedimentation by continuously removing CSF with absorbent paper. His technique was a decisive stimulant for CSF cytology. JUNKER [81] and ULBRICHT [211] had already suggested investigation of CSF cells under the phase contrast microscope after careful concentration of the specimen through centrifugation or sedimentation. Under certain circumstances this technique, introduced by ZERNICKE [236], even permits the observation of well-differentiated cytoplasmic and nuclear structures of a large number of living cells [86]. A considerable disadvantage of the procedure is the limited durability of the slide preparation; photographic documentation is necessary. SEAL [184] utilized the membrane filter, which had previously been used in ultrafiltration, for the first time in cytodiagnosis. Cells can be quickly separated from fluid through the use of suitable filters. This process soon found wide acceptance, especially in English-speaking countries. An additional method of isolating cells from specimens of low concentration was developed by SIMON and SCHROEER [186], who made use of a synthetic fibrin clot. Each of the techniques mentioned above has its own advantages and disadvantages. A more detailed description and evaluation of each process follows.

Sedimentation Technique of SAYK*

SAYK's sedimentation technique represents an improvement on the spontaneous sedimentation process of SCHÖNENBERG. Cells are allowed to settle in a glass or plexiglass cylinder with a diameter of 10–12 mm and a height varying from 40–80 mm. The open-ended hollow cylinder is pressed on a mounting plate by one of several possible mechanisms (lever, weight, or spring, according to author) in such a manner that the system can be regulated. A glass slide placed on the mounting plate forms the actual bottom of the cylinder, and a piece of absorbent paper with a hole identical with the inner diameter of the cylinder is placed between the two glass components. The cells sink to the bottom of the cylinder (the glass slide) under the influences of gravity and of the viscosity of the medium, while fluid containing few or no cells is continuously removed from the cylinder by the absorbent paper. The rate of fluid removal is determined by the pressure of the cylinder on its base and is usually regulated so that sedimentation is completed in 30–40 minutes, at most in 1 hour. The sediment is then ready for fixation.

Slides prepared with this method demonstrate excellent preservation of most cells and permit the use of all staining procedures [105]. The quality of the slides is comparable to that of blood or bone marrow smears. However, the applicability of the method is limited by two serious disadvantages. At least 30% of the available cells are absorbed by the filter paper at the bottom of the cylinder and are lost for the purpose of diagnosis. Some authors [22, 35] estimate cell loss at more than 70%. The loss primarily involves small cells, i.e., lymphocytes, and this must be taken into account in quantitative analysis of CSF. In the examination of more highly concentrated cell preparations, such losses are correspondingly less important. Cell loss can be limited to a certain extent if the absorbent paper is moistened with normal saline before the CSF is placed in the cylinder. An additional disadvantage of SAYK's method is that the fluid absorbed by the paper is not available for further analysis.

There are numerous variations on SAYK's principle of forced spontaneous sedimentation. BAMMER [11] used a chamber in which the pressure of the cylinder on the paper, slide and mounting plate was adjusted by adding or removing weights. KOLAR [101] described an apparatus similar to SAYK's, but capable of preparing several slides at once. The following figures (p. 4) show the sedimentation chamber constructed by the author [100], in which cylinder pressure can be regulated by means of a spiral spring. The device is simple in construction and operation. In addition to the above-mentioned authors. BISCHOFF [19], BOTS *et al.* [22], DEN HARTOG JAGER [35], OLISCHER [138], and WIECZOREK [223] all use the sedimentation technique of SAYK.

Alternate Sedimentation Techniques

Alternate sedimentation techniques were developed in order to eliminate the principal disadvantages of SAYK's method, which are loss of

* Manufacturer: Dr. Günter Lange K.G., Feinmechanische und elektrische Meßgeräte. DDR 1193 Berlin-Treptow.

CSF and loss of a portion of the cells. ENESTRÖM's [47] method is notable in that cell-free CSF is removed from the cylinder by means of a suction device at the surface of the liquid rather than with absorbent paper at the base. A specimen of CSF containing cells is allowed to stand in a primary chamber, where the majority of the cells settle on a slide. The fluid then passes from the bottom of the primary chamber through a capillary fissure into a secondary chamber and is subsequently drawn into a burette. The cells which have managed to pass into the secondary chamber are allowed to settle. Several slides can be prepared from a single specimen, and the cell-free fluid contained in the burette can be used for further analysis. Cell losses are minimal with this method. According to ENESTRÖM's reports, 80–90% of the cells in the specimen are to be found in good condition on the slide.

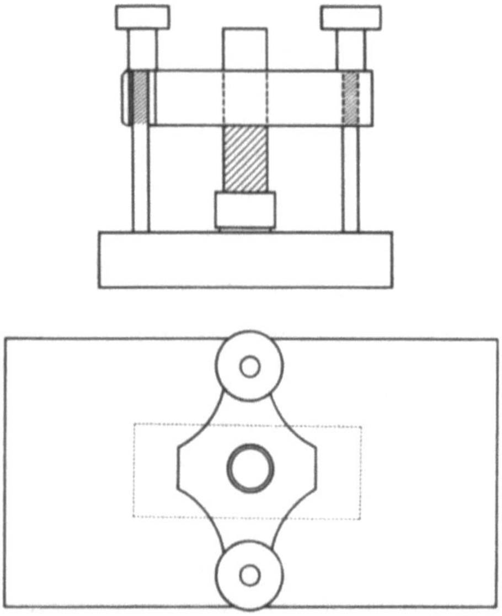

Cell Catch Procedure

This method of increasing cell concentration was first presented by SIMON and SCHROEER [186, 187]. The technique is based on the principle of coagulation, whereby cellular and liquid components of the blood are separated. The cells in a specimen of CSF become entangled in an artificial fibrin network which is produced

The figures show our modification of SAYK's sedimentation chamber.

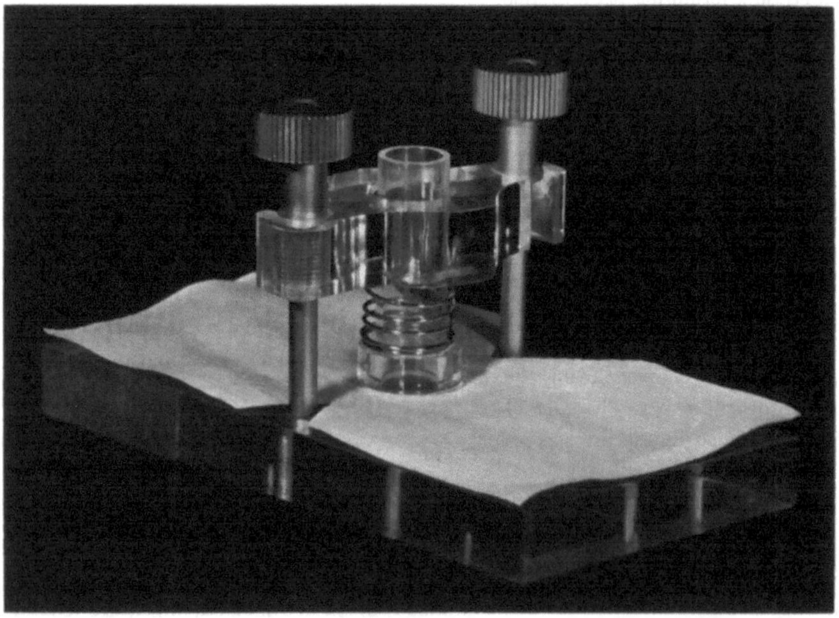

by adding fibrinogen and thrombin to the specimen. The fibrin clot containing the cells is placed on a glass slide which rests on a specially constructed support. The cleared fluid is able to run off, while the temporarily gelatinous fibrin clot is pressed out by means of a second, siliconized glass slide. The air-dried preparation accepts all staining techniques. One succeeds in collecting almost all the cells, but they show obvious morphologic changes. In addition, the fibrin network absorbs the chosen stain to a certain extent. Finally, the cells do not lie in the same plane, a condition which reduces the quality of the photographic image. Judged on the basis of these considerations, the method seems to be inferior to the filtration techniques described below.

Filtration Techniques

Seal [184, 185] suggested filtration of liquid through semipermeable filters, usually consisting of cellulose esters, in order to attain higher cell concentration [14, 16, 56, 92, 122, 123]. The CSF is forced through a membrane filter by means of suction or pressure, which ought not to exceed 15–20 mm Hg. In order to be capable of retaining cells, the filter should be approximately 0.15 mm thick and should contain pores no wider than 5 μ. The diameter of the filter can vary with the apparatus in which it is employed. As a rule, these filtration systems are rather simple in construction. All the cells from the specimen are retained on the filter and made available for cytologic evaluation, if their morphologic condition permits it. The cells can be fixed on the membrane shortly after collection of the specimen, and the cell-free CSF can be submitted to further analysis. A disadvantage of the filters in general use is that they are not resistant to a number of chemical agents, especially methyl alcohol, which cause them either to swell or to dissolve completely. This characteristic of the filters required the development of special staining techniques, whose quality is unfortunately not equal to that of the standard methods. The Papanicolaou stain had to be modified as well. A further consideration is the fact that the cells entangled in the filter network take on a globular shape. This inhibits examination of fine detail in the cytoplasm and renders comparison with cells in standard smear preparations especially difficult. In addition, the membrane filter absorbs stain which must be removed in subsequent decoloration baths, a process which is only partially successful and which in turn impairs the quality of the preparation. Finally, the cells on the filter lie in different planes, which considerably diminishes photographic quality.

In recent years, membrane filters have been developed (polycarbonate filters), which have the advantage of being resistant to many chemical agents. The filters can also be rendered transparent, which allows the use of such common staining techniques as those of Wright and Pappenheim without modification. Cell preservation on these filters is better than on ordinary cellulose ester filters, but one still has to accept some of the limitations described above [61, 93, 158, 160].

Centrifugation

It has already been mentioned that the standard technique—centrifugation of the CSF in a conical tube followed by smear preparation of the sediment—produces unsatisfactory cytologic results. However, special centrifuges have been developed which offer many advantages over conventional models and which are capable of isolating well-preserved cells. The cells are centrifuged at low speed and thrown directly onto a glass slide. Cell-free CSF can be removed simultaneously by filter paper [24, 104, 217] or drained off by means of special accessories [220]. With the former method, the CSF is lost for the purpose of additional analysis, while with the latter, it can be used further.

In the past few years, cytocentrifuges from Shandon Scientific Compagny* have been used

* Shandon Scientific Compagny Limited, 65 Pound Lane, London NW 10 (Great Britain).

increasingly in CSF cytology [39, 71, 234]. The rotor head of this centrifuge contains 12 chambers, each of which receives 2 ml of CSF. At a rotation speed of 800 to 1200 rpm, cells are thrown onto microscope slides, and fluid is retained on absorbent paper, as is also the case in SAYK's method. 12 preparations can be completed in 10 min, or even less, if the CSF is so rich in cells that less fluid is put in the chambers. There is less cell loss with this method than with SAYK's spontaneous sedimentation. On the other hand, centrifugal force causes morphologic changes, especially in the cytoplasm. In our experience, the staining properties also change through centrifugation. Finally, this elegant method requires a rather elaborate technical installation.

Phase Contrast and Interference Phase Microscopy

These refined microscopic techniques offer a variety of possibilities. The methods were originally developed to permit observation and evaluation of cells which had neither been fixed nor stained—in short, living cells. In CSF cytology, a drop of concentrated fluid is placed on a slide and covered with a coverslip, which is sealed with paraffin. Stained or unstained cell preparations—whether concentrated by spontaneous sedimentation or centrifugation—can also be observed under the phase contrast microscope [81, 86, 175, 188, 211].
Phase contrast microscopy is particularly suitable for observation of living cells and for demonstration of cytoplasmic detail, the nuclear membrane and nucleoli. Simple phase contrast microscopy has the disadvantage that cell structures cannot be brought into focus in their entirety; constant micrometer adjustment is necessary. In addition, photographic documentation does not result in pictures of optimal quality. NOMARSKI [132] eliminated almost all of these disadvantages with the introduction of interference contrast microscopy. This technique provides a sharp image of all cytologic details without the so-called halo effect of the phase contrast microscope. The optical impression of three-dimensionality opens interesting new cytologic perspectives. However, the observer must always recognize that three-demensionality results from differences in optical density and not from actual differences in the depth of cellular structures.

If one were to ask which method of cell concentration is most suited to CSF cytology, one would certainly receive a number of different answers from the researchers involved. The membrane filter offers advantages, if one can manage its difficult staining properties, an achievement which requires careful and continuous effort. The researcher who values cytomorphologic detail is better served by the cytocentrifuge or, even more, by SAYK's sedimentation method. We tend to favor this last technique. The sedimentation chamber is easy to use and guarantees optimal cytologic preparation. Slides can be stained with any method, and stain can be removed and reapplied. The phase contrast microscope can be employed without a disruptive refracting medium around the cells.

1.2. Staining Methods [115, 139]

May-Grünwald-Giemsa Stain

Panoptic staining after Pappenheim is the standard technique in hematology, and it has proved to be the most suitable routine method in CSF cytology [27].

Procedure

1. Air-dry the preparation. (Cells that have been fixed or stained while still moist retain their three-dimensional form. This makes comparison more difficult.)
2. Place the slide in a horizontal position and cover with May-Grünwald stain (alcoholic solution of eosin and methylene blue) for 3 min.
3. Pour off staining solution and rinse quickly with tap water.
4. Cover the slide with freshly filtered Giemsa solution (1 part standard solution: 9 parts tap water) for 15 min.
5. Pour off staining solution and rinse quickly with tap water.
6. Place preparation in a vertical position and allow to dry thoroughly.
7. Cover with a standard, nonacidic sealer.

With this stain, the cytoplasm appears in varying blue tones, the nucleus in shades of violet. Nuclear und cytoplasmic structures are distinct. If the nucleus is, for some reason, very heavily stained, renewed rinsing in tap water will lighten the color and reveal the nucleoli and nuclear membrane with greater clarity. The water used in preparing the Giemsa solution and in rinsing must be neutral, as acidic or alkaline water changes the staining characteristics of the nucleus and cytoplasm. Tap water proved to be most suitable for our work, but—if it does not fulfill these specifications—it must be replaced with buffer solution. Pappenheim staining is unsuccessful if the preparation is made with a standard filter, because this type is not resistant to the methyl alcohol contained in the May-Grünwald solution.

Staining the Membrane Filter

Kistler *et al.* [92] recommend the following technique for staining cells that are concentrated with standard membrane filters in routine diagnosis:

1. Preparation of standard solutions:
   - (a) methylene blue 75 mg
     glycerin 25 ml
     phenol 5% 100 ml
   - (b) parafuchsin 5 g
     ethyl alcohol 95% 95 ml
   - (c) Add 1 part of (b) to 9 parts phenol 5%.
2. Preparation of working solution:
   solution (a) 2.5 ml
   solution (c) 5.5 ml
   aqua bidest. 10.0 ml
3. Attach membrane filter to glass slide (cell side up) with metal clips and let stand 3–5 min in working solution.
4. Submerge filter successively in aqua dest., ethyl alcohol 70%, ethyl alcohol 95% and absolute ethyl alcohol (30 sec each).
5. Agitate slightly in isopropyl alcohol and isopropyl alcohol-xylol (2 min each).
6. Submerge filter in xylol or immersion oil. (Both render the filter transparent.)

This staining technique allows an acceptable cell differentiation but is inferior to the Pappenheim technique, because differences in blue and violet tones are less distinct. In addition, stain residue in the filter diminishes the quality of the photographic image.

Papanicolaou's Stain

The staining method introduced by Papanicolaou has proved especially useful in tumor cytology, and it can provide additional information in CSF cytology, under certain conditions. The wide staining spectrum for nuclei and cytoplasm produced with this technique is especially impressive, but the method is too complicated for use in routine diagnosis.

Procedure

1. Submerge the preparation in
   80%, 70%, 60%, 50% methyl alcohol
   for 3 min each.
2. Rinse with distilled water.
3. Let stand in filtered Harris' hematoxylin solution for 6 min.
4. Rinse quickly with distilled water.
5. Submerge the preparation six times in 3% HCL-alcohol solution for short periods.
6. Submerge in
   50%, 60%, 70%, 80%, 96% methyl alcohol
   for 3 min each.
7. Let stand in orange G6 solution for 30 sec to 1 min.
8. Rinse twice with absolute alcohol.
9. Let stand in polychrome EA 31 solution for 30 sec.
10. Submerge twice in 96% alcohol.
11. Submerge twice in absolute alcohol.
12. Rinse in xylol and seal immediately (Eukitt, DePX, Permount, for example).

The technique is not suitable when standard filters are used for cell concentration. In this case, the manufacturers suggest certain modifications.

Nitroblue Tetrazolium (NBT) Reaction

Procedure

1. Prepare 0.2% stain with physiologic saline.
2. Working NBT solution: dilute (1:1) 0.2% NBT stain with phosphate buffer (pH 7.2).
3. Mix 1 ml CSF with 1 ml working NBT solution and with 70 units heparin and incubate for 15 min at 37° C.
4. Make membrane filter preparation after 15 min at room temperature or allow to settle for 15 min in sedimentation chamber after Sayk.
5. Air-dry.
6. Stain after Pappenheim.

Only granulocytes that show a distinct, coarse-grained, dark violet deposit (formazan) in the cytoplasm can be judged positive.

In bacterial meningitis, the CSF can often only be analyzed when the disease has passed into the proliferative or the chronic stage. Distinguishing between the cellularity characteristic of bacterial meningitis and that of viral or tuberculous meningitis can be extremely difficult. The same problem affects clinical differential diagnosis. The total cell count is usually below 1,000/mm^3 and does not allow conclusions as to the etiology of the condition, while the differential cell count shows a mixture of granulocytes, large and small lymphocytes, plasma cells and histiocytes. Bacteria are no longer evident, since almost all have been destroyed through phagocytosis or antibiotic therapy. Bacterial cultures can be sterile under certain conditions. The nitroblue tetrazolium test (NBT test) seems to be useful in such situations.

Park [147] first used the test on granulocytes in bacterial infections, but it also gives a positive granulocyte reaction in the acute, subchronic, and chronic phases of bacterial meningitides [34, 54]. Yellow NBT pigment is reduced to blue-violet NBT by enzymes (dehydrogenases) which develop in the cytoplasm of the granulocyte in the course of the phagocytosis of bacteria [120]. Reduced pigment is deposited in the cytoplasm in the form of large granules or flakes, and the proportion of cells containing the pigment is then determined. If more than 12% of the granulocytes react positively, one may conclude that a bacterial infection has occurred. Mycosis also gives a positive reaction in the NBT test, while viral meningitis reacts negatively. Tuberculous meningitis probably gives a negative reaction as well.

Ferrous Staining (Prussian Blue Reaction)

Procedure

1. Air-dry.
2. Submerge in absolute alcohol
   90% methyl alcohol
   80% methyl alcohol
   70% methyl alcohol
   for 3 min each.
3. Rinse with distilled water.
4. Let stand in potassium ferrocyanide solution (1 part 2% HCL:1 part 2% potassium ferrocyanide) for 25 min.
5. Rinse with distilled water.
6. Let stand in potassium ferrocyanide solution for 20 min.
7. Rinse with distilled water.
8. Submerge in 70% methyl alcohol
   80% methyl alcohol
   90% methyl alcohol
   absolute alcohol
   for 3 min each.
9. Seal the dry preparation.

Cytoplasmic structures containing iron stain gray-blue to brilliant blue with this technique [146]. These cytoplasmic components can be spread over the entire cytoplasm or deposited in discrete patches. Melanin does not react this way, so differentiation is possible.

Lipid Staining with Sudan Black or Sudan Red

Procedure

1. Air-dry.
2. Fix the preparation by submerging it twice in 50% methyl alcohol for short periods.
3. Let stand in filtered Sudan black or Sudan red for 20 min.
4. Submerge quickly in 50% methyl alcohol.
5. Rinse quickly with tap water.
6. Let stand in modified Harris' hematoxylin for 10 min in order to stain nuclei.
7. Seal the air-dried preparation with liquified glycerinated gelatin.

The lipid droplets in the cytoplasm are stained black with Sudan black stain, scarlet red with Sudan red.

Fluorescent Staining with Acridine Orange

Procedure

1. Submerge preparation in
   95%, 80%, 70%, 60%, 50% methyl alcohol for 3 min each.
2. Acetic acid for 1 min.
3. Rinse with distilled water.
4. Stain with 0.01% acridine orange phosphate buffer for 3–5 min.
5. Submerge preparation in phosphate buffer for up to several hours, until staining is completed.
6. Differentiate in calcium chloride solution ($CaCl_2$) for 1–2 min.
7. Rinse with phosphate buffer in order to stop differentiation.
8. Cover with coverslip.

In order to prepare the acridine orange phosphate buffer solution (limited stability), dilute 0.1% acridine orange standard solution (unlimited stability) with phosphate buffer (pH 6.0). Phosphate buffer solution: mix 4 parts 1/15 M Na_2HPO_4 (9.465 g in 1,000 ml distilled water) with 23 parts 1/15 M KH_2PO_4 (9.072 g in 1,000 ml distilled water). The resulting solution has pH of 6. 0.1 M $CaCl_2$: dissolve 11.099 g $CaCL_2$ in 1,000 ml distilled water.

It is advisable to store the finished preparation (soaked in phosphate buffer, covered with a coverslip) in a chamber saturated with phosphate buffer. For microscopy, the light source should be rich in ultraviolet rays; high pressure mercury or xenon bulbs are most suitable. An excitatory filter must lie between the lamp and the condenser, an absorbent filter between object and ocular to shield the eye from damaging ultraviolet rays. For the results with this technique, see page 63.

PAS Reaction

Procedure

1. Fix the air-dried preparation in formol vapor for 5 min.
2. Wash in tap water for 10 min.
3. Let stand in 0.8% fresh periodic acid (H_5SO_6) for 8 min.
4. Wash in tap water for 8 min.
5. Let stand in Schiff's reagent for 30 min.
6. Rinse with sulfite solution three times (2 min each) in separate cuvettes.
7. Soak in tap water for 5 min.
8. Stain nuclei with modified Harris' hematoxylin for 8–10 min.
9. Air-dry.

Carbohydrates in the cells are demonstrated with the PAS reaction. Aldehydes are released by periodic acid and are stained with fuchsin, which is contained in Schiff's reagent. One can distinguish homogeneous, fine-grained, and coarse-grained PAS-positive substances in the cytoplasm. Demonstration of coarse-grained substance in the cytoplasm of lymphoblasts is important in differential diagnosis, as is the positive reaction of vacuole secretions in adenocarcinomas.

Demonstration of Nonspecific Esterase with Alpha-naphthyl Acetate [141]

Procedure

1. Fix the air-dried preparation for 5 min in formol vapor.
2. Rinse in water for 5 min and air-dry.
3. (a) Mix 0.25 ml Na-nitrite solution with 0.25 ml pararosaniline solution with 25 ml 0.2 M phosphate buffer (pH 7.0).
   (b) Dissolve 50 mg alpha-naphthyl acetate in 1 ml pure acetone. Allow mixture of (a) and (b) to act on preparation for 30 min.
4. Rinse with water for 5 min.
5. Stain with modified Harris' hematoxylin for 8 min.
6. Soak for 15 min.

Esterase staining should take place within two days of the preparation of the slide. The reaction is brown, either homogeneous or granular. All monocytic and histiocytic cell elements react positively. Various types of leukemia can be differentiated with this test.

Peroxidase Demonstration after Graham-Knoll

Procedure

1. Air-dry the preparation.
2. Fix in formol alcohol (1 part 40% formol and 9 parts 96% alcohol) for 30 min.
3. Rinse in tap water.
4. Let stand in benzidine reagent for 5 min. (Dissolve a pinch of benzidine in 12 cm^3 96% alcohol and dilute with 8 cm^3 water. Add 0.04 ml fresh H_2O_2 solution).
5. Rinse with water.
6. Air-dry.
7. Stain with Giemsa solution for 20 min.

Peroxidase-positive substances—such as granulocytes at all stages of maturity—stain yellow-brown. Myeloblasts may occasionally or temporarily react negatively.

Gram Stain

Several slides should be prepared when clinical evidence leads one to suspect bacterial meningitis. Preparations with Pappenheim stain and gram stain are a necessary minimum, so a CSF cytology laboratory ought to be equipped for these diagnostic procedures.

Procedure

1. Air dry the specimen, and fix in a flame.
2. Immerse in Hucker's crystal violet for $^1/_2$ min.
3. Rinse with Burke's iodine (=Lugol's solution).
4. Immerse in Burke's iodine for $^1/_2$ min.
5. Decolorize with 95% ethanol.
6. Immerse in safranin (1 : 5 dilution) for $^1/_2$ min.
7. Rinse with distilled water. Dry.

Gram-positive bacteria (e.g. streptococci, micrococci) appear dark blue, gram-negative (e.g. Neisseria, Pseudomonas, Enterobacter) red.

2. The Cells of Normal Cerebrospinal Fluid

A specimen of CSF taken in the lumbar area and unaltered by diseases of the central nervous system normally contains two types of cells, i.e., lymphocytic and monocytic forms. These designations are used in recognition of the great similarity between these cells and the corresponding cells of the blood. These are apparently reserve cells of the CSF, which are maintained at a constant level. The ratio of lymphocytes to monocytes is about 70 to 30 [155, 165, 189].

A few granulocytes appear rarely and irregularly in addition to the cell forms described above, while some red blood cells, which find their way into the sample as a result of the puncture, are almost always present.

It must be emphasized that the findings listed above apply only to the first cubic centimeters of the sample. If additional portions are analyzed, both qualitative and quantitative variations in the cells become apparent [204]. The cell count may rise to 20 or even 30/mm^3, while the proportion of monocytic cells increases and that of the lymphocytic forms decreases. More monocytes appear, and cells of the ependyma or of the choroid plexus are occasionally present. One has to take into account the fact that the number of transformed or degenerated monocytes is usually greater in infants and small children than in adults.

2.1. Lymphocytes and Monocytes

Lymphocytes. These are small, relatively isomorphic cells with dark, compact, round to slightly oval nuclei and usually a narrow border of pale-blue cytoplasm. This type can be divided into subcategories on the basis of staining properties and width of the cytoplasmic border. The small to medium-sized light-cytoplasmic lymphocyte predominates in normal CSF; larger or dark-cytoplasmic forms appear only exceptionally.

The origin of the lymphocytic cells in the CSF is unknown. SAYK [165] believed that the majority develops out of leptomeningeal stem cells. A considerable body of evidence indicates that lymphocytes emigrate from the blood to the CSF [183]. It is also possible that the lymphocytes emigrate back into the blood, so that a regular circulation and exchange exist.

Monocytes. One may term these cells "monocytes" of the CSF, when they are similar to those of the blood. We believe that they are inactive cell forms of the 'mononuclear phagocytic system' [108]. Differentiation of these cells often presents difficulties, since definitive morphologic and cytochemical criteria have not been established. This may be the reason why the term "monocyte" ist not universally applied to these cells when they appear in the CSF. The cells are large and have an eccentric, oval, kidney-shaped or horseshoe-shaped, blue-gray nucleus, which often contains large, pale nucleoli. The cytoplasm is pale blue-gray, occasionally interspersed with vacuoles. Under the phase contrast microscope, one observes a lively, rapid but rapidly diminishing modulation of the cytoplasmic outline of the cells in the specimen.

A small proportion of the cells in normal CSF can also be classified as monocytes even though they differ somewhat from the latter. The cells in question have a larger, round or irregular nucleus, which is usually rich in chromatin. The cytoplasm is also more intensely stained. These cells differ from monocytes in their morphology, and probably in their development, though not in their origins. Since these cells have basic characteristics in common with monocytes, they can be classified as "monocytoid".

The monocytic forms normally degenerate more rapidly in vitro than do the lymphocytic cell forms. As in the case of the lymphocytes, some authors believe that the monocytes develop from leptomeningeal stem cells [165], while others suspect a hematogenic origin [136, 151].

*Plate 1**

a–e Lymphocytes. (c) Transformed lymphocyte (above), of a type which does not appear in normal CSF. (e) Interference phase microscopy. × 1,250.

f–m Monocytes. (f, g) Interference phase microscopy. × 1,250. (h) Monocyte mitosis, rarely seen in normal CSF.

---

* Technical notes:

1. All specimens were obtained with a sedimentation technique after SAYK.
2. Unless otherwise noted, Pappenheim stain was used throughout.
3. With few exceptions, magnification was × 400 or × 1000 (both with oil emersion), and originals were enlarged with a constant factor.

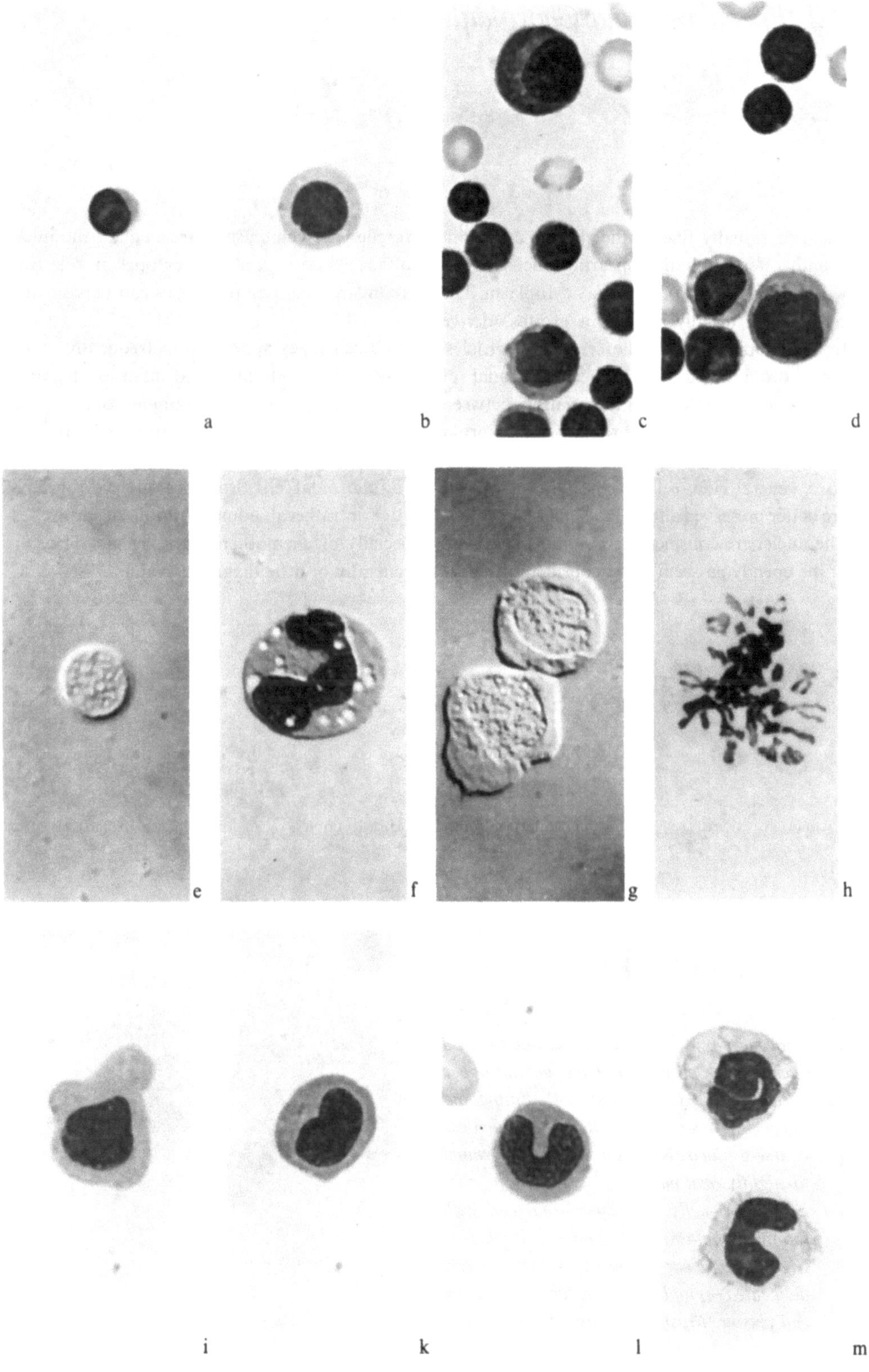
a
b
c
d
e
f
g
h
i
k
l
m

2.2. Ependymal and Choroidal Cells

One occasionally finds rather fragile cells with round, often pyknotic-appearing nuclei and a wide border of pink or blue-gray cytoplasm. The cells, which usually appear in groups, derive from the epithelium of the cerebral ventricles, the choroid plexus, or the central canal of the myelon. Cytologic differentiation between cells of the ependyma and those of the choroid plexus is usually difficult.

As a general rule, cells of the choroid plexus are better preserved and appear in larger groups. The nuclei are conspicuously isomorphic. Cells of the ependyma seem to be considerably more fragile. The cell clusters are smaller, the nuclei often pyknotic and the cytoplasm relatively abundant. Staining properties can vary greatly.

Both cell types appear more frequently in the CSF of small children and in cases of hydrocephalus as well as in samples obtained by cisternal or ventricular puncture or by pneumencephalography [76, 204, 229]. Ependymal and choroidal cells also appear in the cytogramm after intrathecal administration of drugs, especially of chemotherapeutic agents. They are generally of little diagnostic value.

Plate 2

a–c Choroid plexus cells. (a) Plexus cell with 3 nuclei. (b) Cell cluster. Cilia on the upper border of the cytoplasma. (d) Epithelioid cluster of several cells with wide, indented, coarse-grained cytoplasm and uniform, round to slightly oval nuclei.

d–f Ependymal cells. (d) Ependymal cell with foamy cytoplasm. (e) Loose cell cluster. Signs of nuclear pyknosis. Wide, vaguely delineated cytoplasmic border. (f) Loose cell cluster. Most nuclei are pyknotic.

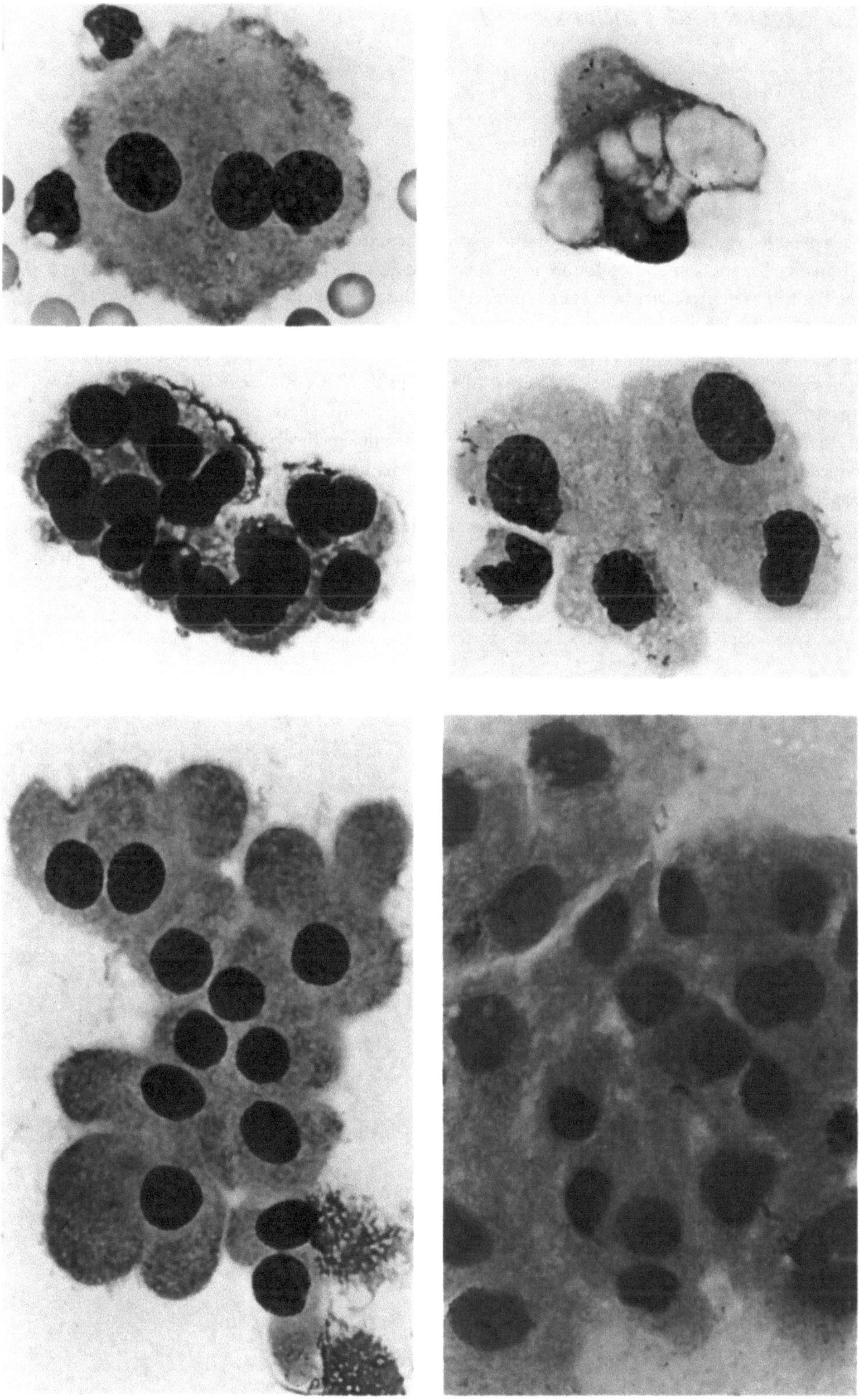

a d b e c f

2.3. Accidental Findings

Cells which originate in the path of the aspiration needle are sometimes found in a specimen. They are often cartilage cells, unmistakable in their typical appearance. Ganglion cells turn up rarely [46], most often after ventricular pneumencephalography, sometimes appearing in cell clusters and sometimes singly. In the latter case, the cytoplasm is sharply reduced, and the cell seems to consist of a bare nucleus with a very large nucleolus. Capillaries appear rarely in the preparation, usually as a consequence of neurosurgical intervention.

One sometimes finds immature cells of the hematopoietic system in the CSF of small children and cachectic, old patients, even though leukemia is not present [194]. These cells enter the CSF as a result of an injury to the marrow of a vertebra caused by the puncture needle. This may cause initial confusion, but the simultaneous presence of cells of the myelopoietic, erythropoietic, and thrombopoietic systems usually indicates the source of this condition.

Impurities in the cell picture generally consist of cork fragments, cellulose fibers, bacteria, or fungus spores.

Plate 3

a Capillary vessel (ventricular pneumencephalography in porencephaly).

b Cartilage cell cluster.

c Artificial mixture of cells of the hemopoietic system. Megakaryocyte and several stages of granulocytopoesis (promyelocyte, myelocyte, stab cell).

d Ganglion cells and endothelium of blood vessels. Large, centrally-located nucleoli (ventricular pneumencephalography).

e Ganglion cell cluster. Nissl stain allows better demonstration of cytoplasmic components.

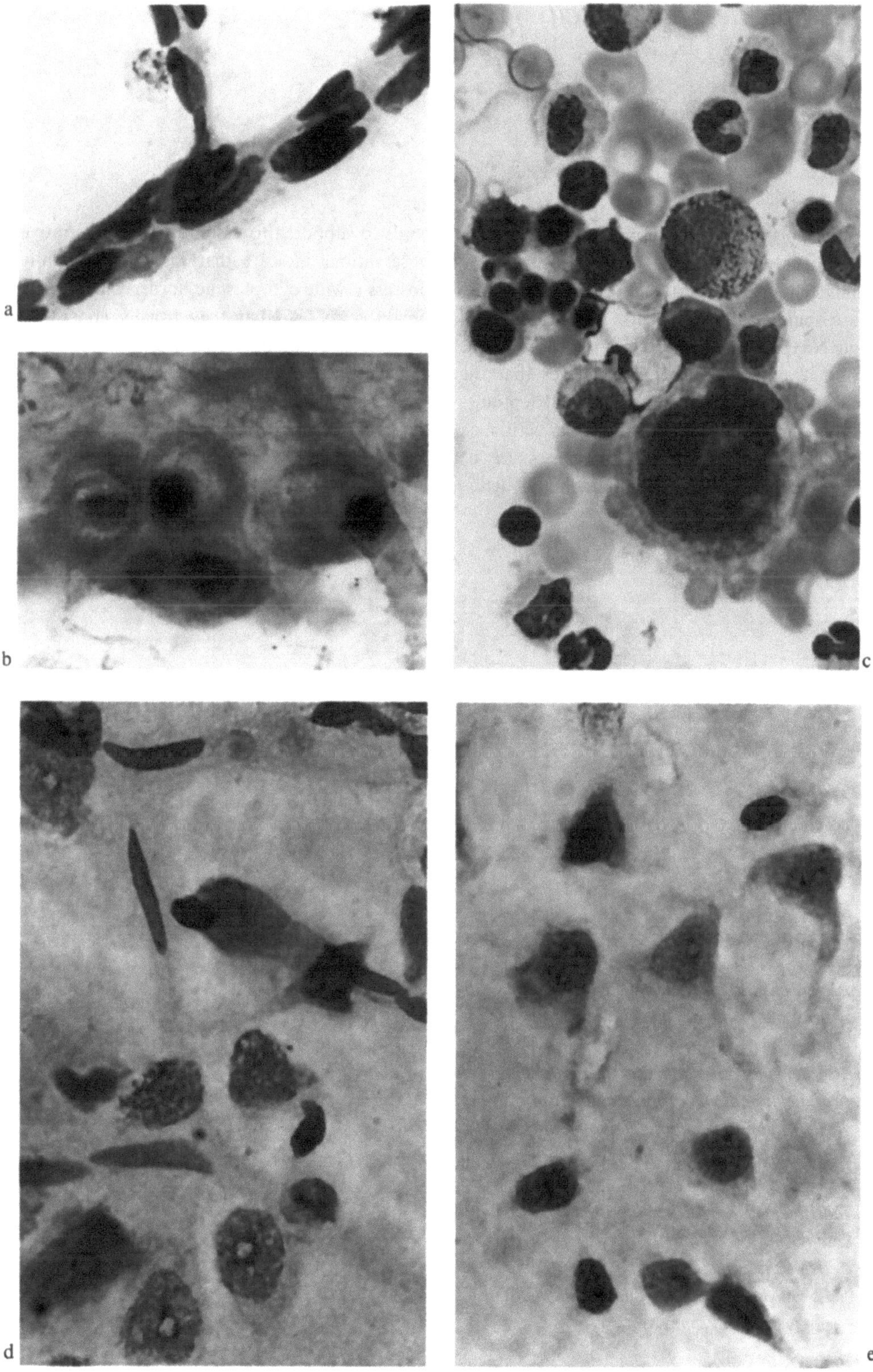
a
b
c
d
e

2.4. Spindle-Shaped Cells

In the portions of a sample obtained through pneumencephalography, one almost always finds cells which, as a rule, consist only of a bare spindle-shaped nucleus. ECKES *et al.* [45] and NAYLOR [129] have already reported this finding, and OLISCHER [138] believes that the cells are derived from the superficial leptomeningeal cell layer. The fact that cells with such spindle-shaped nuclei are sometimes to be found with undifferentiated arachnoid tissue tends to support this assumption. Comparative observations indicate that the cells originally possess a wide cytoplasmic border which easily breaks away, so that they usually consist of nothing more than a bare nucleus when they arrive at the lumbar area. They have no special diagnostic significance, but may indicate that the CSF specimen was obtained by pneumencephalography or ventricular puncture.

Plate 4

a Loose cluster of heterogeneous cells. Cells with spindle-shaped nuclei above and below right.

b Leptomeningeal cell similar to epithelium with two nuclei and wide diffuse cytoplasm between clusters of undifferentiated leptomeningeal cells.

c Cell, probably same origin as 'a', with pigment deposit in the cytoplasm.

d Individual cell with obvious tendency to shed cytoplasm.

e Cells of the same origin consisting of bare nuclei, which are oval or spindle-shaped (after pneumencephalography).

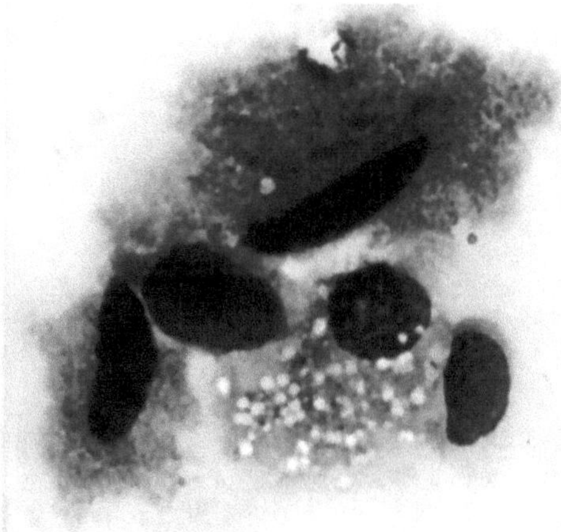
a

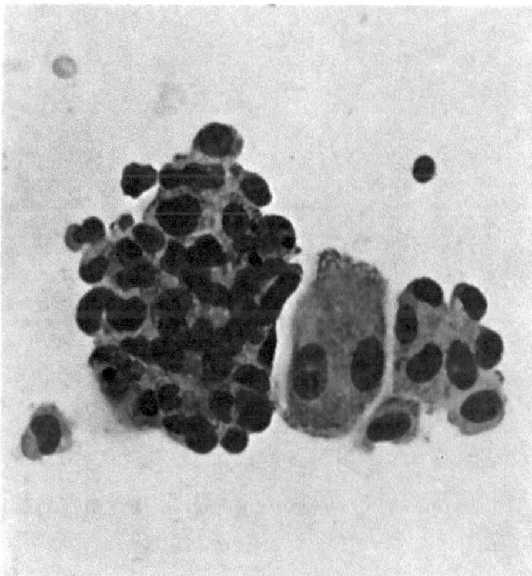
b

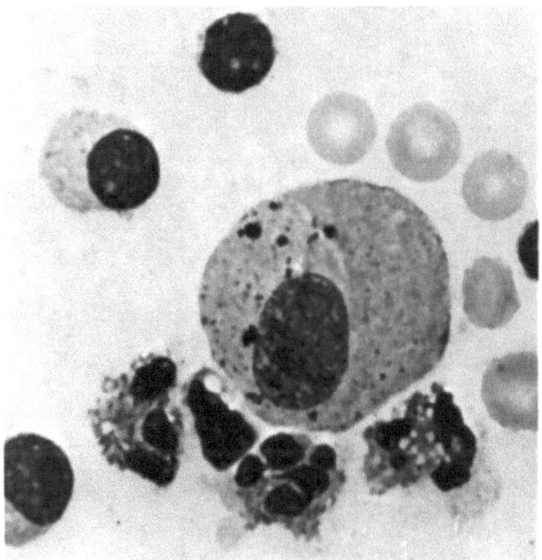
c

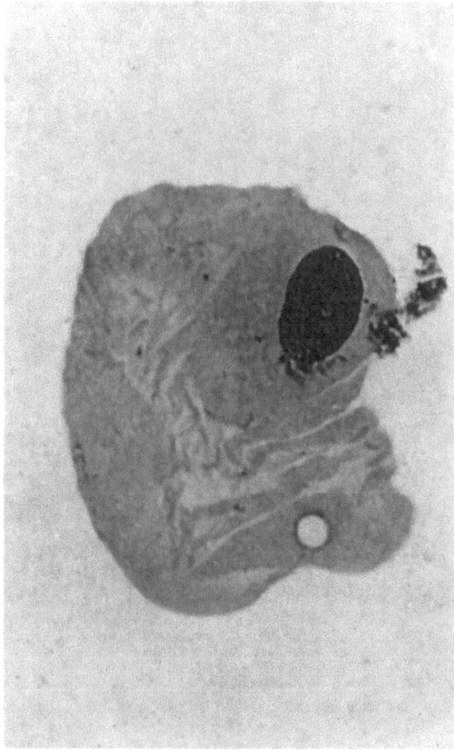
d

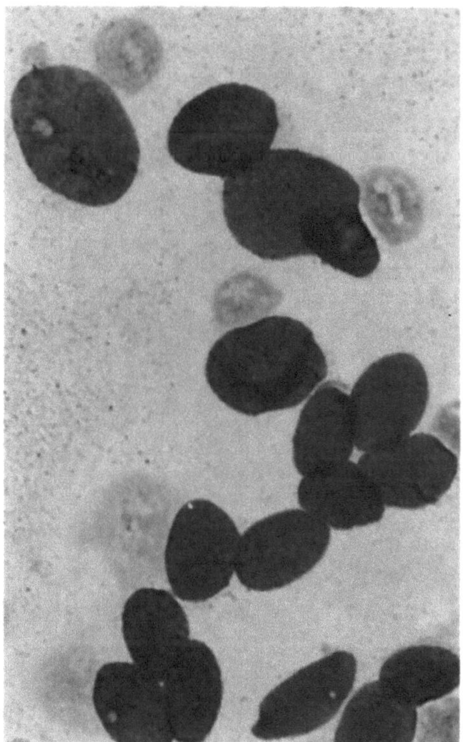
e

3. Cells of the Mononuclear Phagocytic System

3.1. Development of Macrophages

The term macrophage, in the sense of a particular type of cell in the CSF, is not universal. We define these cells of the Mononuclear Phagocytic System [108] according to their origin, morphology, cytochemical properties and their ability to phagocytize corpuscular substances of various sizes. The designation may therefore include histiocytes, as well as endothelial cells, reticulocytes and cells that develop from microglia [23, 36, 59, 67, 75, 165, 188, 207].

The macrophages develop for the most part from pluripotential cells in the reticuloendothelial tissue of the leptomeninges. It is probable, as SAYK believed, that the macrophages as well as the lymphocytic and plasmocytic cell forms can develop from these leptomeningeal stem cells. In fact, the young macrophages which appear more often in infants and small children with subarachnoid hemorrhage cannot be distinguished morphologically from large lymphocytes or plasma cells. These promonocytes are often organized in loose groups, have large, round, centrally-placed nuclei and deep-blue cytoplasm with Pappenheim stain. It is probable that the cytologic determination of the macrophages is already completed at the proliferation stage of the leptomeningeal parent tissue; the apparently undifferentiated and pluripotential freshly exfoliated promonocytes are incapable of developing into other cell types. Despite the unfavorable conditions of the CSF, the macrophages retain their vitality for a relatively long period after detaching themselves, singly or in groups, from the parent tissue. Continued mitotic and phagocytic activity is evidence of this capacity for survival. The vitality of the cells apparently increases in proportion to the pathologic conditions which brought about their activation.

In general, two types of macrophages are distinguishable in the CSF. The first type is similar to the monocyte and comprises 2% of the cellular component of normal CSF [189].

Although some authors believe that the cells originate in the blood [151], the assumption is questionable. Perhaps the cells are a special survival form of exfoliated leptomeningeal promonocytes. It may be that these cells migrated at an early stage from the blood into the leptomeninges. One could term these temporarily settled cells 'histiocytes'. The monocytic forms described above are capable of phagocytosis but are generally not active as phagocytes. If large-scale phagocytosis is necessary, the promonocytes are transformed into macrophages and perform the function. The cytoplasm of these cells can vary and is sometimes strongly basophilic, the nucleus round to oval, though sometimes even segmented. The nucleus can be either centrally or excentrically located, and one or two nucleoli are often present. The cells described here usually appear singly, but can be found in rather

Plate 5

a–c Clusters of undifferentiated leptomeningeal cells.

d Unusually strong phagocytic cell reaction. Four days after ischemic cerebral stroke.

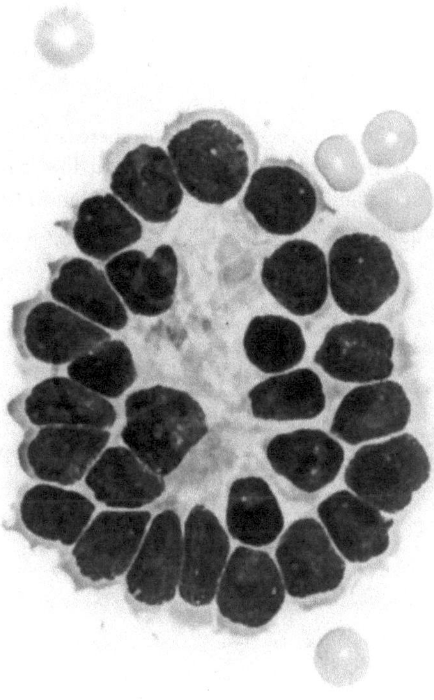
a

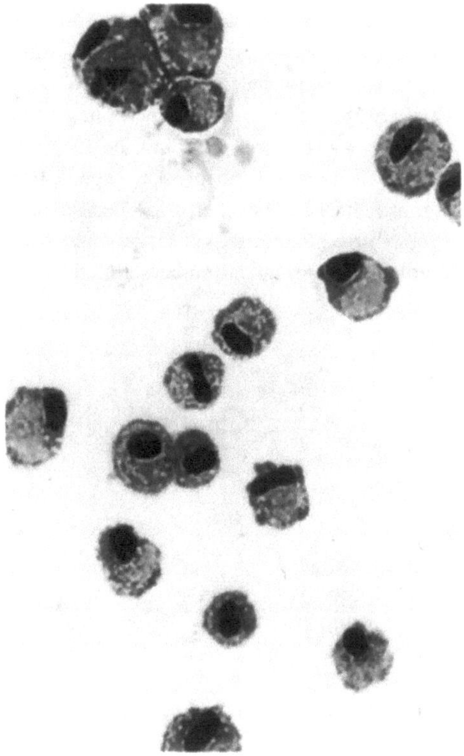
d

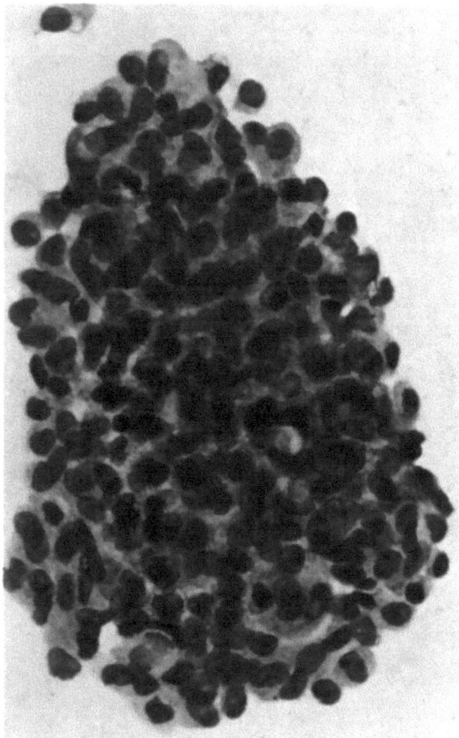
b

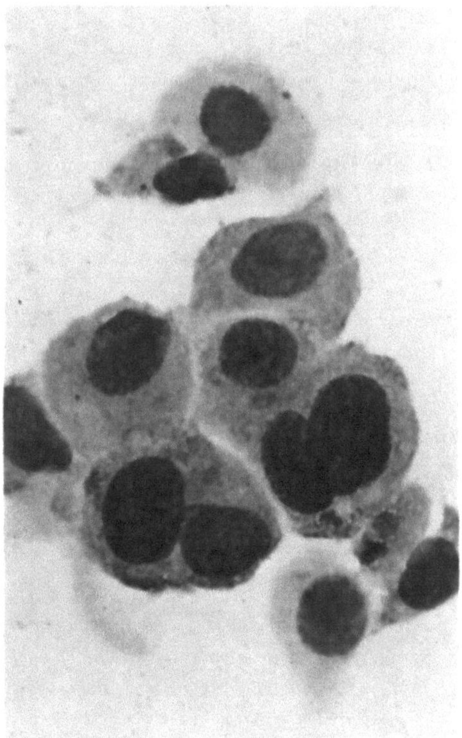
c

large groups, especially when the activating stimulus is of recent origin. Macrophages transformed into giant cells with several or many nuclei are then not a rarity. Macrophages survive days or even weeks and can be identified as signs of a pathologic process; hence their diagnostic value. They perform specific tasks in the course of infectious disorders and occasionally appear in large numbers after cerebral stroke [190]. Their primary function is the elimination of foreign matter by means of phagocytosis. The death of the phagocyte is characterized by nuclear pyknosis, enlargement and granulation of the nucleoli and increased fraying of the vacuolated cytoplasm. At this stage, the cells and their ingested material can in turn be phagocytized by younger macrophages.

Plate 6

a, b Patient: female, 4 weeks old. Cerebral trauma at delivery. CSF: 120 cells/mm³. Loose, undifferentiated leptomeningeal cell clusters.

c, d Same patient: 2 weeks later. CSF: 430 cells/mm³, 120 mg% protein. The leptomeningeal stem cells have developed for the most part into macrophages, some of which show degenerative stigmata.

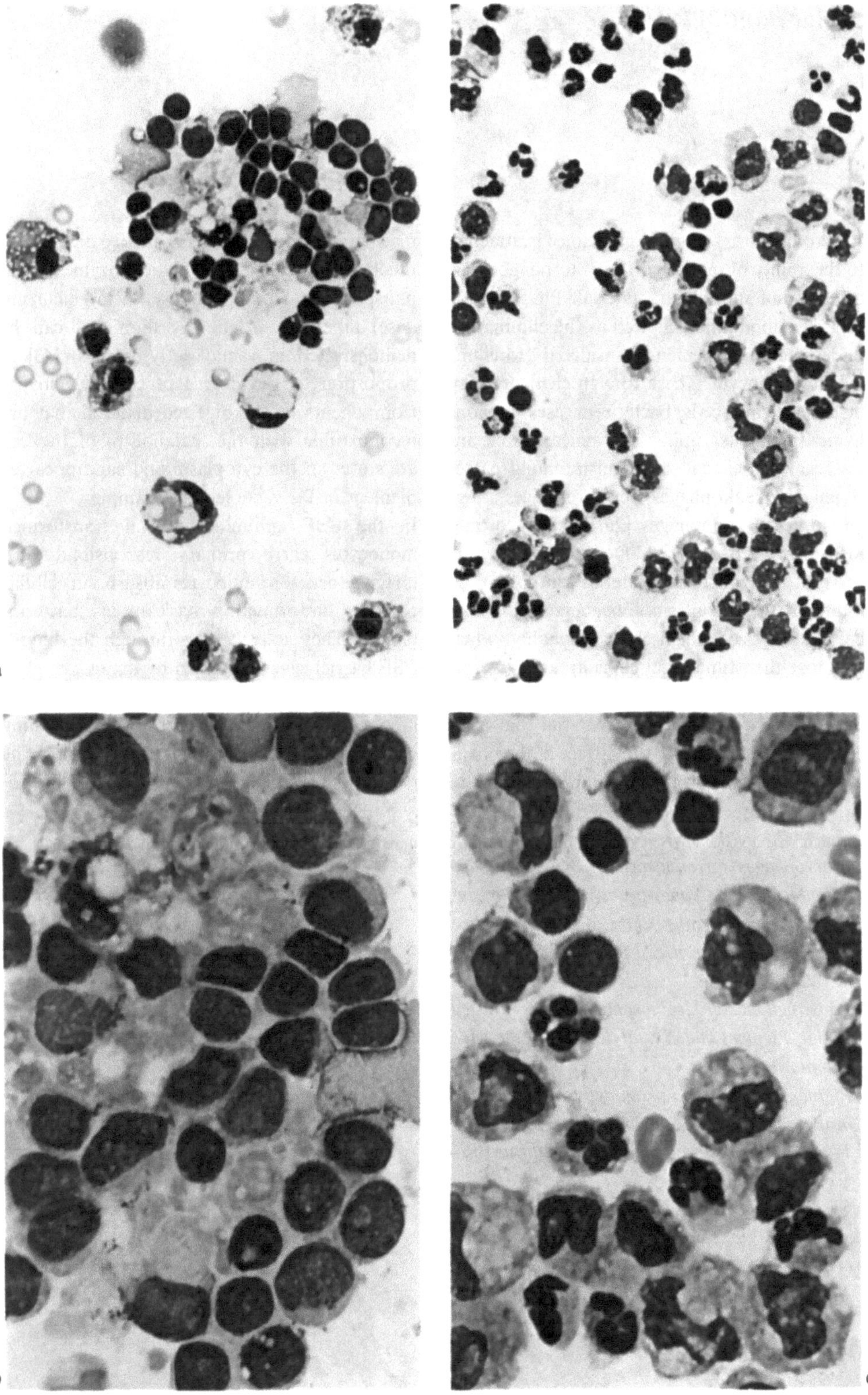

a c b d

3.2. Phagocytosis

In cytodiagnosis, special significance is attached to the ability of many cell forms to phagocytize cellular and subcellular material. Phagocytosis can serve food intake as well as the elimination or destruction of elements considered "foreign" by the phagocyte. These foreign elements may be either living (cells, bacteria, viruses) or non-living (pigments, lipids, substances given intrathecally). One can distinguish among bacteriophages, leukophages, erythrophages, pigmentophages, lipophages [26], etc., according to the ingested material. Deposits of various sorts of phagocytized material can often be found in the cytoplasm. Phagocytes can also be phagocytized themselves, probably when they lose their functional capacity and take on the character of foreign elements.

In the presence of globulin, material that is to be phagocytized clings to the surface of the phagocyte, is then surrounded and enclosed by plasma pseudopodia, and is finally digested as much as possible. Hydrolytic enzymes (acid phosphatase, aminopeptidase, beta-glucuronidase) take part in the digestion and can be demonstrated in vacuoles (lysosomes) [43] in proportion to the degree of cellular activity. Components of the digested material that are incompatible with the metabolism of the cell are stored in the cytoplasm and can appear as pigment or fat vacuoles, for example.

In the CSF, granulocytes and transformed monocytes carry primary responsibility for phagocytosis. Granulocytes absorb subcellular particles and function basically as bacteriophages. They actively pass through the blood-CSF barrier when attracted by toxins.

The phagocytic activity of the transformed monocyte is extraordinary; it absorbs anything which acts on it as a foreign body, even if the particle is several times larger than the phago-

Plate 7

a Macrophages surrounded by RBC's. Irregular cytoplasmic landscape, probably a sign of increased activity rather than of a degenerative process. Interference phase microscopy. × 1,250

b Loose cell cluster. Isomorphic, undifferentiated nuclei. Pigment and crystal deposits in the wide cytoplasmic border.

c Pigment deposits and phagocytized lymphocyte in the cytoplasm of a macrophage.

d Phagocytosis of foreign bodies after brain surgery.

e Several lymphocytic cells grouped around two phagocytes. A few phagocytized cells in the cytoplasm.

f Macrophage completely filled with phagocytized lymphocytic cells, of which the nuclei are pyknotic.

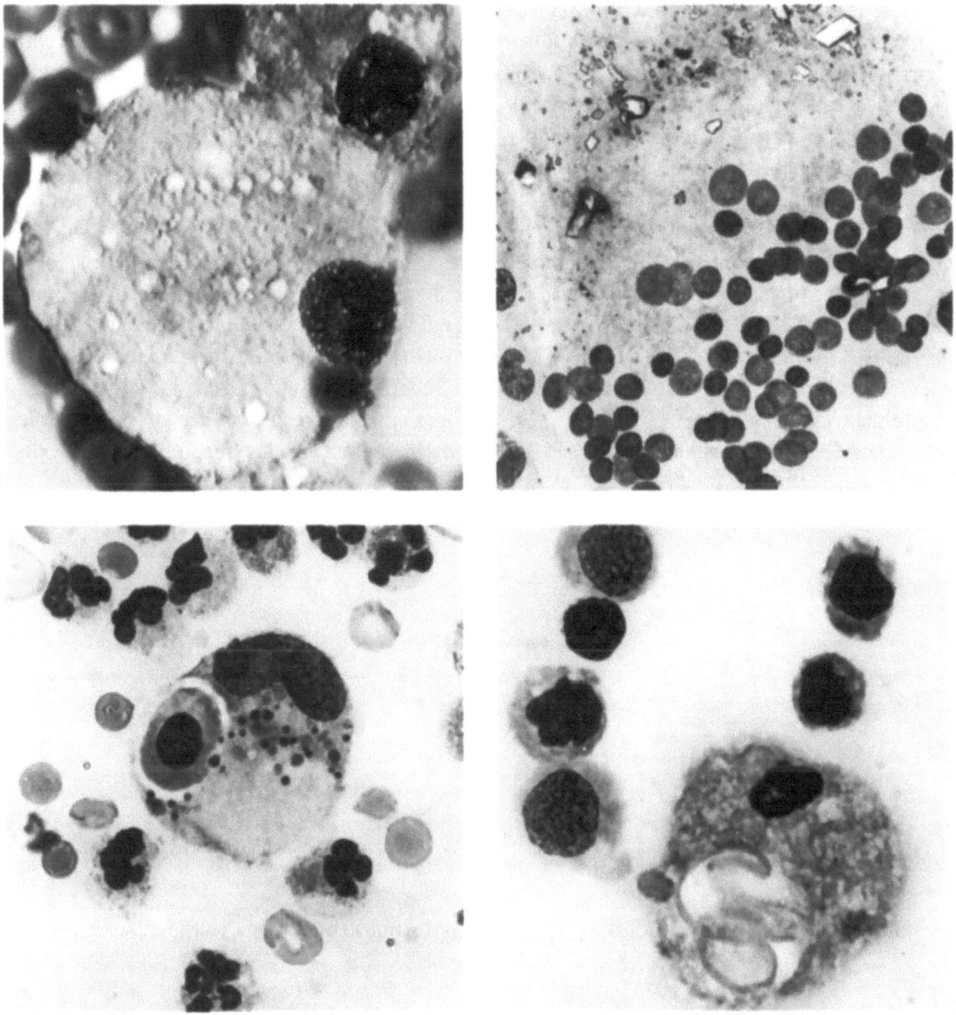
a
b
c
d
e
f

cyte itself, as is often the case [66, 179]. Material which has been phagocytized and broken down by enzymes and which cannot be utilized by the cell is stored in the cytoplasm, where it often takes the form of apparently empty vacuoles. The vacuoles can intersperse the entire cell body with fine or large globules and are characteristic of the so-called gitter cell, a functional state of the macrophage. The boundaries of the vacuoles can tear open and flow together, forcing the plasma and the nucleus to the periphery. The resultant phagocytic form is the signet ring cell, which may attain a diameter of more than 150 μ.

The presence of cytoplasmic vacuoles after panoptic staining does not necessarily indicate that phagocytosis has taken place. Apparently empty vacuoles can just as well be interpreted as being stigmata of degeneration and, therefore, evidence of a cytoplasmic separation process. One speaks of "masked" phagocytosis [165] if cytoplasmic inclusions do not allow a definite judgment as to whether phagocytosis has taken place, or if the nature of the phagocytized material cannot be identified. The term should be avoided if a diagnosis cannot be confirmed through the use of special stains such as Sudan red or Sudan black [68]. The designation is acceptable if the size of the vacuoles indicates recent phagocytosis of red blood cells.

Plate 8

a Foamy cytoplasm after phagocytosis of oily contrast medium.

b Phagocytosis of melanin in cerebral melanoma metastases. Pigment deposits indistinguishable from hemosiderin with Pappenheim stain.

c Aseptic meningitis after cerebral contusion. Lipid storage cell.

d Sudan red stain. Demonstration of lipids in the cytoplasm of granulocytes as well as masses of lipid in the cytoplasm of phagocytes.

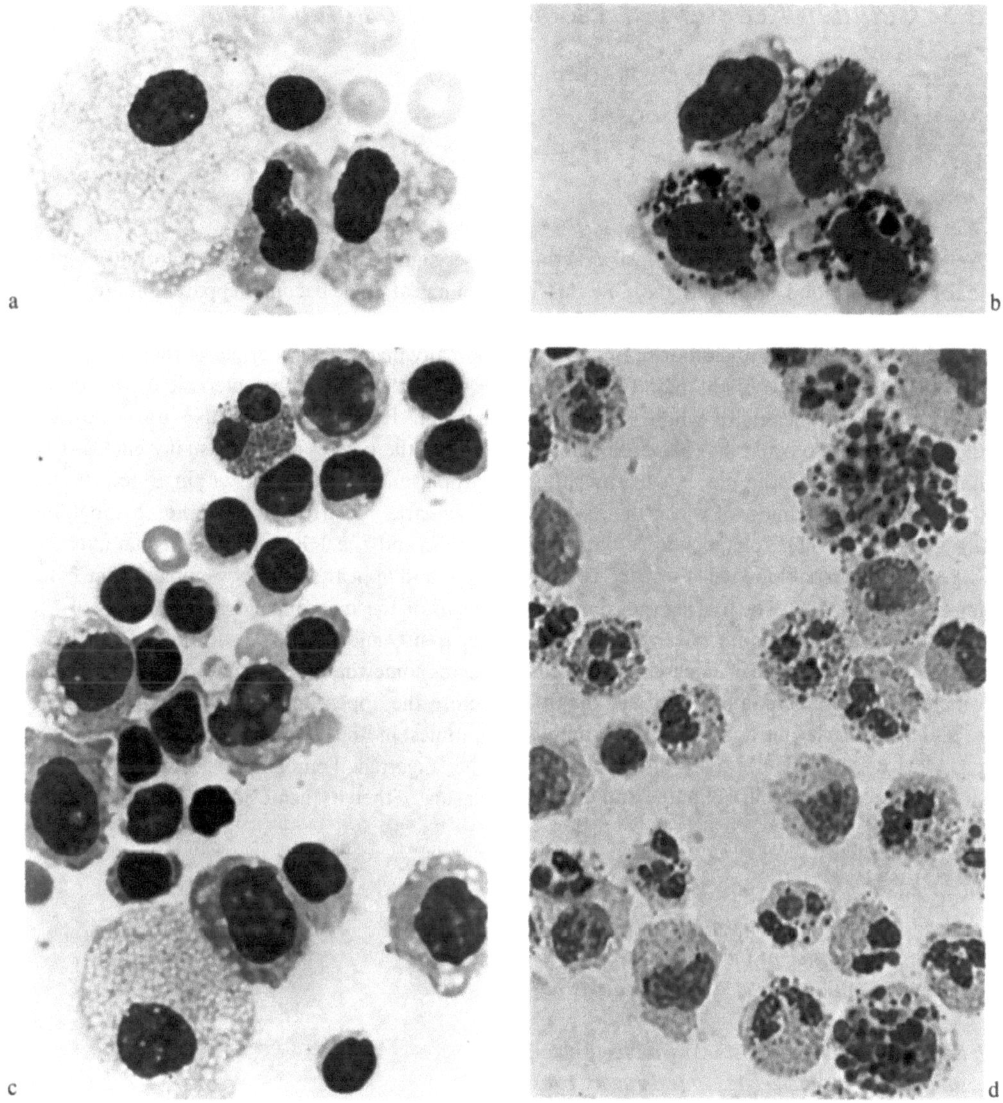

3.3. Cellular Reaction of Bloody Cerebrospinal Fluid

An extraordinary leptomeningeal cell reaction takes place if blood infiltrates the CSF, as is the case after aneurysmal or hypertonal hemorrhage, after cerebral contusion or brain surgery [50, 212]. The presence of the blood as foreign material causes a meningitis whose first sign is the appearance of all the hematologic cell forms. In rare cases, the pleocytosis can attain levels of 1,500 cells/mm^3, of which the majority are granulocytes. The phagocytic activity provoked by the red blood cells (RBC) is most impressive [135, 223]. The first signs of phagocytosis appear about 4 hours after the introduction of blood into the CSF. The monocytic cells, which usually circulate in the CSF, are the first to undergo cytoplasmic activation. They initiate phagocytosis, but their activity in the elimination of the red blood cells is limited and quickly loses significance. Lively proliferation of leptomeningeal tissue becomes apparent about 12–18 hours after the initial stimulus. The appearance of many macrophages, some of them organized in loose clusters, is characteristic of this development. Red blood cells take up a position around the phages and are usually enclosed by cytoplasmic pseudopods within a few hours. Enzymatic destruction of the hemoglobin begins, and the RBCs lose their characteristic color and appear as empty vacuoles in the cytoplasm of the macrophages. Destruction of the apparently more resistant stroma of the RBC begins somewhat later, and about 4 days elapse before the appearance of the first hemosiderin granules in the cytoplasm of the phagocyte. The digested hemoglobin is stored in the plasma, either as hemosiderin, which contains

Plate 9

a Blood which has penetrated into the CSF space. Proliferation of the mononuclear phagocytic system. Interference phase microscopy. ×1,250.

b Start of RBC phagocytosis. Macrophage shows similarities to monocyte. Interference phase microscopy. ×1,250.

c Cytoplasm filled with phagocytized RBC's, which retain their normal color. Further proliferation of the mononuclear phagocytic system. Interference phase microscopy. ×1,250.

d Phagocytized RBC's almost completely bleached. Stroma still in evidence. Interference phase microscopy. ×1,250.

e First sign of pigment deposition. Large number of fine vacuoles. Interference phase microscopy. ×1,250.

f Macrophage with 3 nuclei. Conspicuous nucleoli. Rich pigment deposits in cytoplasm. ×1,000.

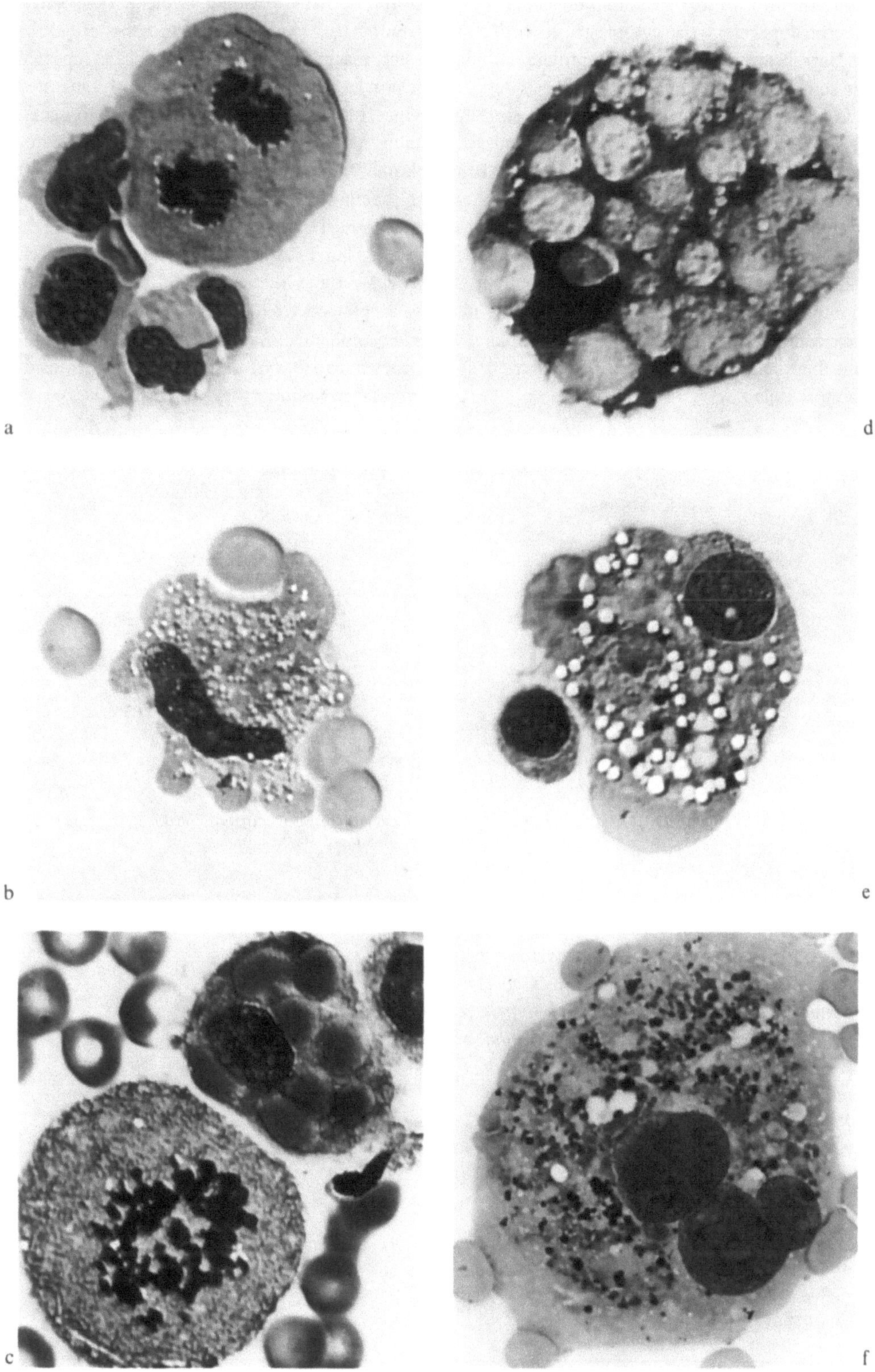
a
d
b
e
c
f

iron, or as iron-free hematoidin. The hemosiderin appears in the cytoplasm in the form of dark-brown to gray-black granules, which are sometimes quite large. Hematoidin is deposited in the cytoplasm somewhat later, usually as brownish yellow granules or as brilliant red crystals. Isolated crystals of hematoidin can sometimes be found in the CSF when the macrophages degenerate. The life span of the phagocytes is relatively long, and they can still be found in the subarachnoid space more than 6 months after the initial hemorrhage. A long life span can be simulated when aging macrophages and their contents are in turn phagocytized by younger cells.

In addition to providing evidence of a single hemorrhage, even several months after the event, macrophages can also indicate repeated hemorrhages. In this case, recently ingested RBCs, bleached RBCs, hemosiderin and hematoidin are simultaneously present in the cytoplasm. Staining with Prussian blue provides the only definitive evidence of hemosiderin in the cytoplasm. The hematogenous pigments take on various shades of blue with this stain, which provides the only certain differentiation between phagocytized melanin and hemosiderin. In melanoma metastases of the CNS, the phagocytes are capable of depositing both hemosiderin and melanin in the cytoplasm.

Plate 10

a Signet ring cell. Nucleus and phagocytized material displaced to the border of the cytoplasm.

b Spinal tap 3 weeks and 2 days after hemorrhage in the subarachnoid space. Demonstration of hemosiderin deposits and fresh phagocytosis of intact red blood cells.

c Phagocyte cannibalism.

d Macrophage with two nuclei and many hematoidin crystals in the cytoplasm, 7 weeks after subarachnoid hemorrhage.

e Demonstration of hemosiderin in the cytoplasm 3 months after cerebral contusion.

f Identification of hemosiderin in cytoplasm of a macrophage. Prussian blue stain.

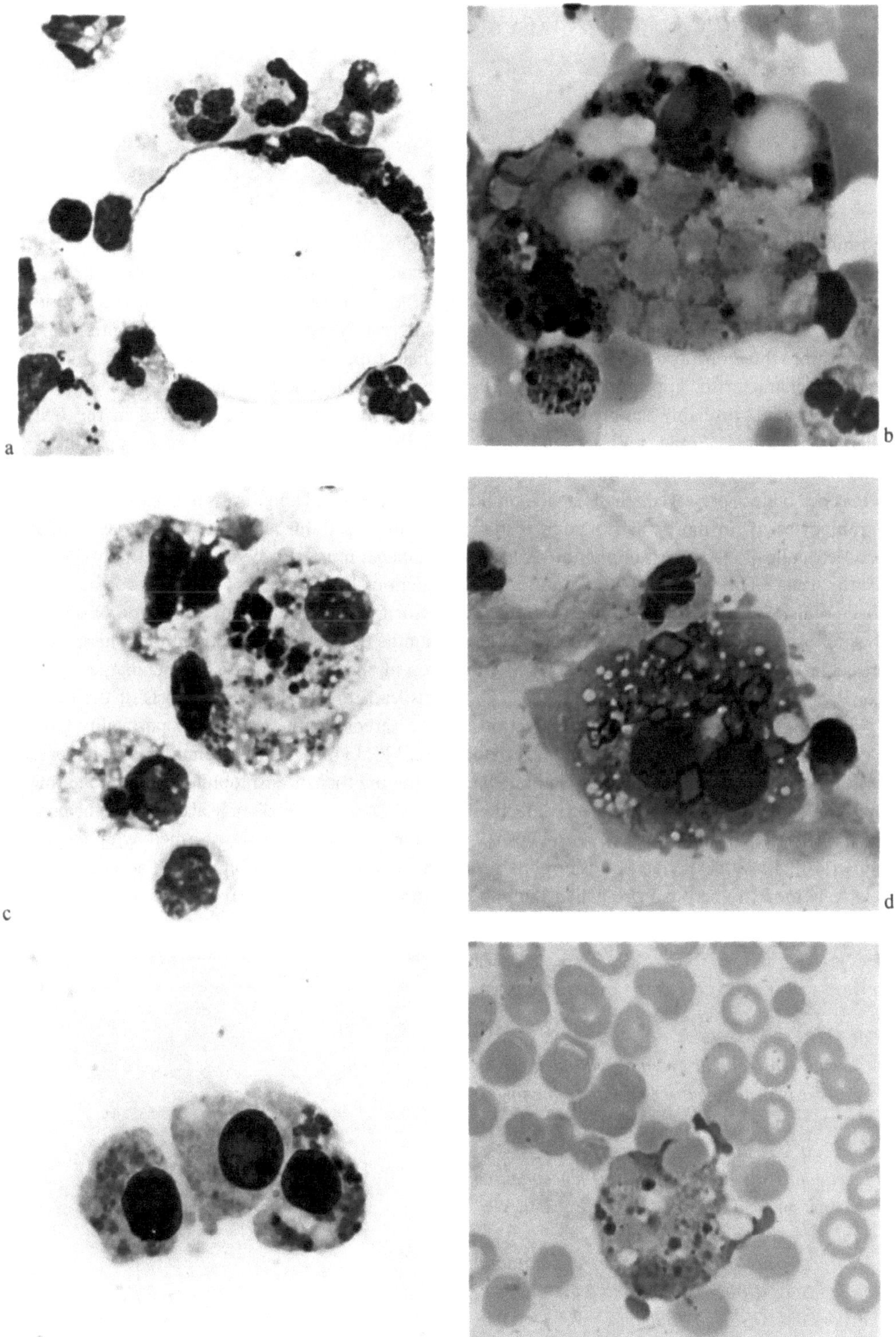

a b c d e f

4. Eosinophilic Granulocytes

Eosinophils can easily be recognized in panoptically stained slide preparations on the basis of their relatively large eosinophilic plasma granules. The nucleus, which often has two, more rarely three or more segments, also gives the cell a characteristic appearance. Eosinophils have a special, probably suppressive, role in allergic processes. Some authors emphasize their vulnerability in vitro [165], but they can often be better preserved than neutrophilic granulocytes, if no more than 3 hours elapse between collection of the sample and fixation of the cells. The normal CSF contains no eosinophils. Since they also appear rarely in the majority of infectious reactions, any increase in their number demands special attention.

Eosinophils can be present in any infectious pleocytosis, but they usually constitute less than 1% of the total number of cells. It is well known that one stage of cellular reaction to infection is characterized by the appearance of eosinophils [170]. In our experience, eosinophils are relatively more numerous in so-called lymphocytic meningitis than in bacterial meningitis. They appear most often in the cellularity that accompanies tuberculous meningitis, although not in the entire course of the disease. Eosinophils rarely predominate in meningitis [233]. Their appearance must evidently be attributed to a specific allergic reaction rather than to a microbe. They can be observed in the CSF in the course of urticaria or of an allergic bronchial asthma. The appearance of aseptic eosinophilic meningitis, which can sometimes be diagnosed through the use of ventricular drainage, must also be attributed to an allergic reaction [163].

Evidence of eosinophils over a period of several months demands special attention. This condition in chronic meningitis or meningoencephalitis leads one to suspect zoonosis of the CNS. Cysticercosis is the most likely possibility [102, 114, 126, 134, 153, 210]. A simultaneous increase in the number of eosinophils in the circulating blood does not necessarily accompany the condition. In the CSF, transformed lymphocytes and plasma cells predominate. The cell count can vary between normal values and more than 1,000/mm^3.

Plate 11

a Chronic eosinophilic meningoencephalitis in cysticercosis of the brain and meninges. Patient: male, 40 years old. CSF: 60 cells/mm^3, 20% eosinophils.

b Eosinophils in tuberculous meningoencephalitis. Patient: male, 25 years old. CSF: 250 cells/mm^3, 6% eosinophils.

c "Foreign body" meningitis after implantation of a Spitz-Holter drain. Patient: female, 3 years old. CSF: 600 cells/mm^3, 90% eosinophils.

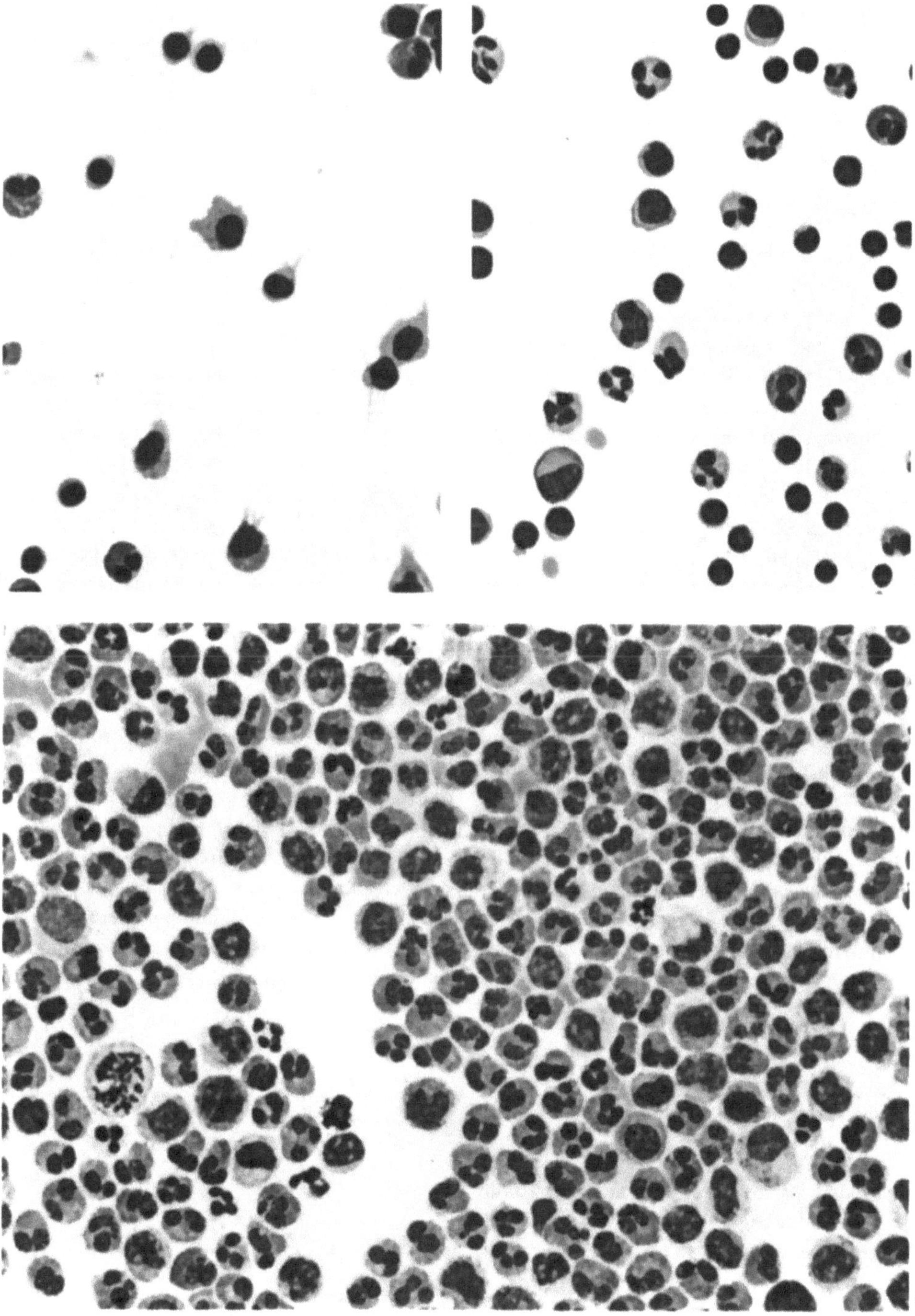

a b c

5. *Plasma Cells*

Plasma cells are present only in pathologically altered CSF.

A number of researchers have suggested that some of them develop from leptomeningeal stem cells [55, 106, 138, 192]. A smaller fraction may emigrate from the blood. It is probable that they develop from small inactive lymphocytes which have emigrated from the blood to the CSF space.

Immature plasma cells usually have a large, centrally-located nucleus and light-blue cytoplasm, while increasingly basophilic cytoplasm is characteristic of maturing cells. Mature cells are conspicuous for their bright blue cytoplasm with panoptic stain. The nucleus of the mature cell is smaller than that of the immature forms, is excentrically-located and contains a grainy chromatin structure that sometimes assumes the typical spoked-wheel form. Plasma cells with two nuclei are more frequently observed in viral diseases and chronic infectious disorders, though the presence of more than two nucleoli is rare. Nucleoli are almost never encountered, although mitoses, often with rather thick chromosomes, are not at all a rarity. A crescent-shaped area of light cytoplasm around the nucleus is typical but not present in every case.

The cytoplasm of the plasma cell is capable of producing and secreting immunoglobulins. Visible evidence of this activity is the appearance of plasma vacuoles, which sometimes become so numerous that the cytoplasm develops a frothy texture. These frothy cells are smaller than those of histiocytic origin, and their vacuoles are more sharply outlined. All of the developmental forms of the plasma cell can be present in an infectious pleocytosis. In this case, cytologic distinction between plasma cells and histiocytes or transformed lymphocytes is often impossible.

Plasma cells are never present in normal CSF; their presence always indicates an inflamed condition of the CNS. They are especially typical of viral diseases and such chronic infections as syphilis, tuberculosis, sarcoidosis, periarteritis nodosa, zoonosis of the CNS, subacute sclerosing panencephalitis and multiple sclerosis [169]. Plasma cells can also be found in a normal cell count in the initial stage of Guillain-Barré

Plate 12

a, b *Mature plasma cells. Large cytoplasmic body and relatively small excentric nucleus. Light perinuclear zone.*

c *Plasma cell with two nuclei.*

d *Specific chromatin structure of the plasma cell nucleus demonstrated with the phase contrast microscope.*

e, f *Rare type of plasma cell. Numerous vacuoles in cytoplasm (mumps meningoencephalitis).*

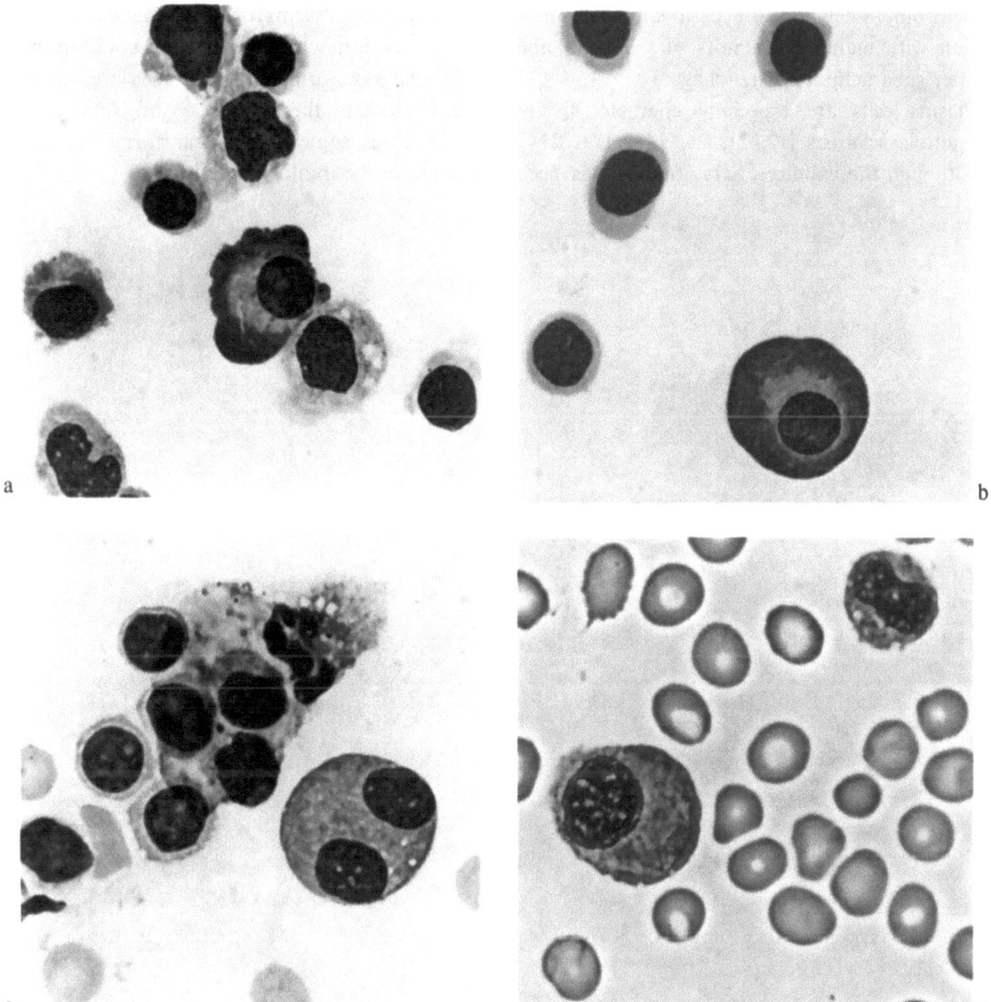
a
b
c
d
e
f

syndrome. They appear occasionally in connection with malignant tumors of the CNS and after subarachnoid hemorrhage.

Plasma cells are especially characteristic of multiple sclerosis [12, 21, 24, 148, 171, 214]. Although the validity of this finding has been disputed [209], it seems to have special diagnostic value in cases in which the total cell count in the CSF remains normal. Plasma cells are sometimes found in these cases, as are numerous lymphocytes, some of them transformed, at the expense of the monocytic cell types.

Plate 13

a–f Various stages of lymphocyte transformation. A few plasma cells. (a) Meningoencephalitis in mumps infection. Patient: female, 2 years old. CSF: 420 cells/mm^3. (b) Meningoencephalitis in tuberculosis. Patient: male, 26 years old. CSF: 260 cells/mm^3. (c, d) First symptoms in multiple sclerosis. (e) Anaphase in the mitosis of a cell similar to a plasma cell (chronic meningitis, etiology unknown). (f) Unusual plasma cell with 6 nuclei. (Aseptic, eosinophilic meningitis after bleeding angioma, Patient 9 years old).

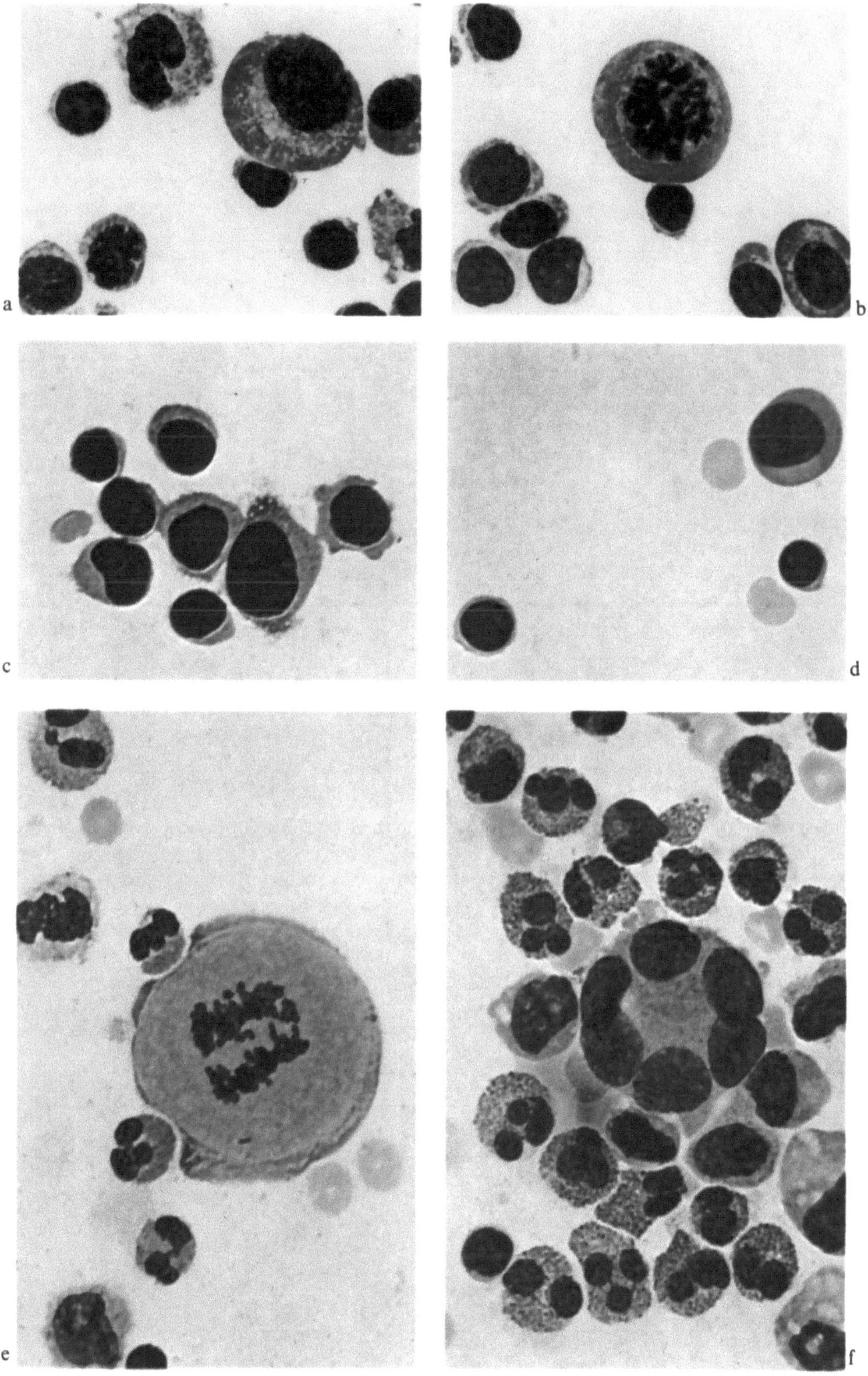

6. *Bacterial Meningitis*

6.1. Acute Exudation Phase

At the onset of bacterial meningitis, the CSF is dominated by the neutrophils, which constitute over 90% of the cells. The granulocytes are attracted by bacterial toxins and penetrate the vascular wall from the blood to the CSF [88]. Their function as microphages is to phagocytize and digest the bacteria. This capability is not unlimited, as aging granulocytes accumulate fat vacuoles and ultimately perish. Degenerated granulocytes become pus, and the CSF can take on a cloudy, yellowish or even creamy, purulent appearance. The cell count can rise to more than 30,000/mm^3. Young stab cells predominate in the early stage of leukocytic invasion, but they are quickly outnumbered by segmented and oversegmented granulocytes. The remaining cells are relatively unaltered monocytes and lymphocytes showing the first signs of plasma activation. The presence of eosinophils should be regarded as a chance finding, while basophils [38] appear occasionally, primarily in small children.

It is advisable to make the cell count immediately after collection of the sample, since granulocytes tend to degenerate if laden with phagocytized material. This tendency can seriously alter the actual cell count and lead to unnecessary diagnostic confusion.

The opportunity of demonstrating bacteria in a CSF preparation during the acute infectious stage is often neglected. Even in panoptically stained preparations, pathogenic organisms can be identified in up to 40% of bacterial meningitides, if care is exercised. It is therefore advisable to prepare a Gram stain in addition to the routine method, if bacterial infection is supected. Bacterial culture and the antibiogram are indispensible, as these techniques offer the clinician valuable information about the characteristics of pathogenic organisms. *Haemophilus influenzae, Escherichia coli* and *Neisseria meningitis* are important pathogens, especially in infants and children, while pneumococci, staphylococci, and streptococci predominate in adults. So-called problematic bacilli (e.g. *Proteus mirabilis, Pseudomonas aeruginosa, Klebsiella pneumoniae*) can cause meningitis for example, after neurosurgical intervention or, rarely, after lumbar puncture.

The bacterial meningitides show no difference in cellularity during the acute infectious stage. The total number of cells alone has no prognostic significance, but a disparity between the amount of bacteria and the constant and limited granulocytic reaction is obviously an unfavorable sign. The granulocytes may continue to predominate for several days after the first application of suitable antibiotic therapy, but the subchronic stage of the infection becomes cytologically evident thereafter.

Plate 14

Bacterial meningitis.

a Pneumococcal meningitis. Patient: female, 65 years old. CSF: 10000 cells/mm^3. Demonstration of predominantly intracellular diplococci. Pure granulocytosis.

b Enterococcal meningitis. Patient: female, 3 years old. CSF: 700 cells/mm^3. Masses of enterococci (primarily diplococci) in an inadequate granulocytic reaction.

c Coli meningitis. Patient: male, 2 years old. CSF: 6000 cells/mm^3. Demonstration of Escherichia coli bacteria in the subacute stage of meningitis.

d Staphylococcal meningomyelitis. Patient: male, 24 years old. CSF: pus, 30000 cells/mm^3. Large clusters of micrococci. Fatty change in granulocytes.

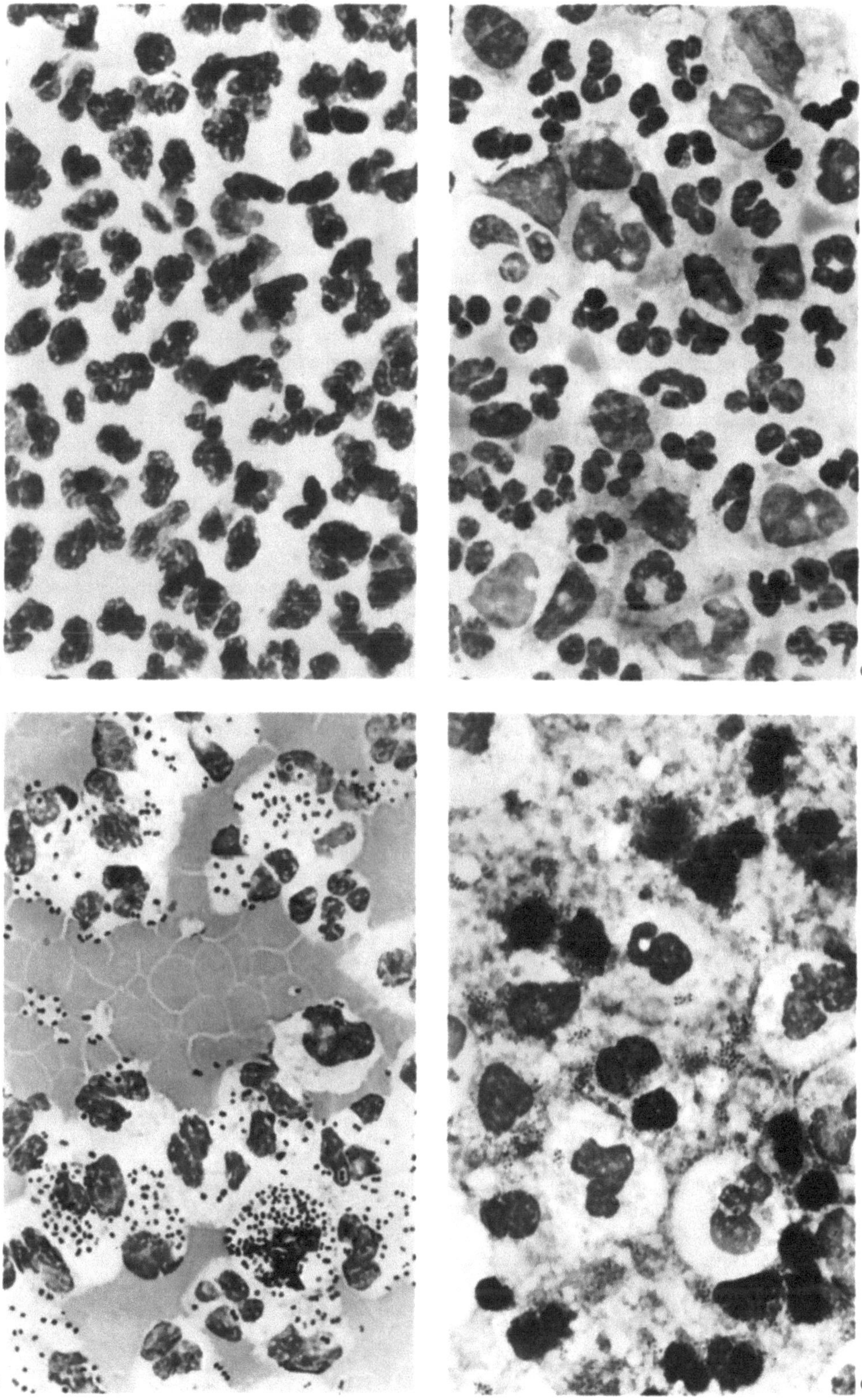

a c

b d

6.2. *Proliferative Phase of Infection and Repair Phase*

Proliferative phase of infection. A rapid decrease in the total number of cells, principally at the expense of the granulocytes, follows the application of effective antibiotic therapy. All of the granulocytes possess degenerative stigmata, their nuclei become oversegmented and pyknotic, and the plasma becomes vacuolated and loses its sharp outline. Activated monocytes become more numerous, and many of them develop into macrophages which initially attack bacteria. In a later stage, they phagocytize granulocytes and plasma cells as well. Macrophages and plasma cells show only limited mitotic activity.

Repair phase. The repair phase generally begins with the almost complete disappearance of the granulocytes, usually after a few days of therapy. The macrophages show signs of aging: their phagocytic activity diminishes, both the nucleus and cytoplasm are sharply indented, and the cytoplasm is interspersed with vacuoles. The plasma cells become numerically less important. The cell count shows first signs of normalization with an increase in the number of less active, small lymphocytes.

The stages of bacterial meningitides described above may be modified by such known factors as the virulence of the pathogenic organism, the resistance of the patient, and the effectiveness of the antibiotic. The proliferative phase can lead either to a renewed outbreak of inflammation, which is always indicated by another increase in the number of neutrophils, or to a chronic stage, rather than to a healing phase. The cytologic characteristic of the chronic stage is the almost equal proportion of activated monocytes, lymphocytes and granulocytes. In

Plate 15

a, b *Fatal mixed infection after neurosurgical intervention. Patient: male, 16 years old. CSF: 42 cells/mm³.*
Dense accumulation of bacteria in the granulocytes. Inadequate granulocytic reaction.

c *Pneumococcal meningitis. Patient: female, 28 years old. CSF: 1100 cells/mm³, 85 mg% protein. Lumbar puncture 8 hours after development of headache.*
Masses of typical, lance-shaped diplococci, some in chains. Purely granulocytic cell reaction. Some stab cells.

d *Early proliferative phase of pneumococcal meningitis. Start of the monocytic reaction. Patient: male, 60 years old. CSF: 800 cells/mm³, 120 mg% protein.*

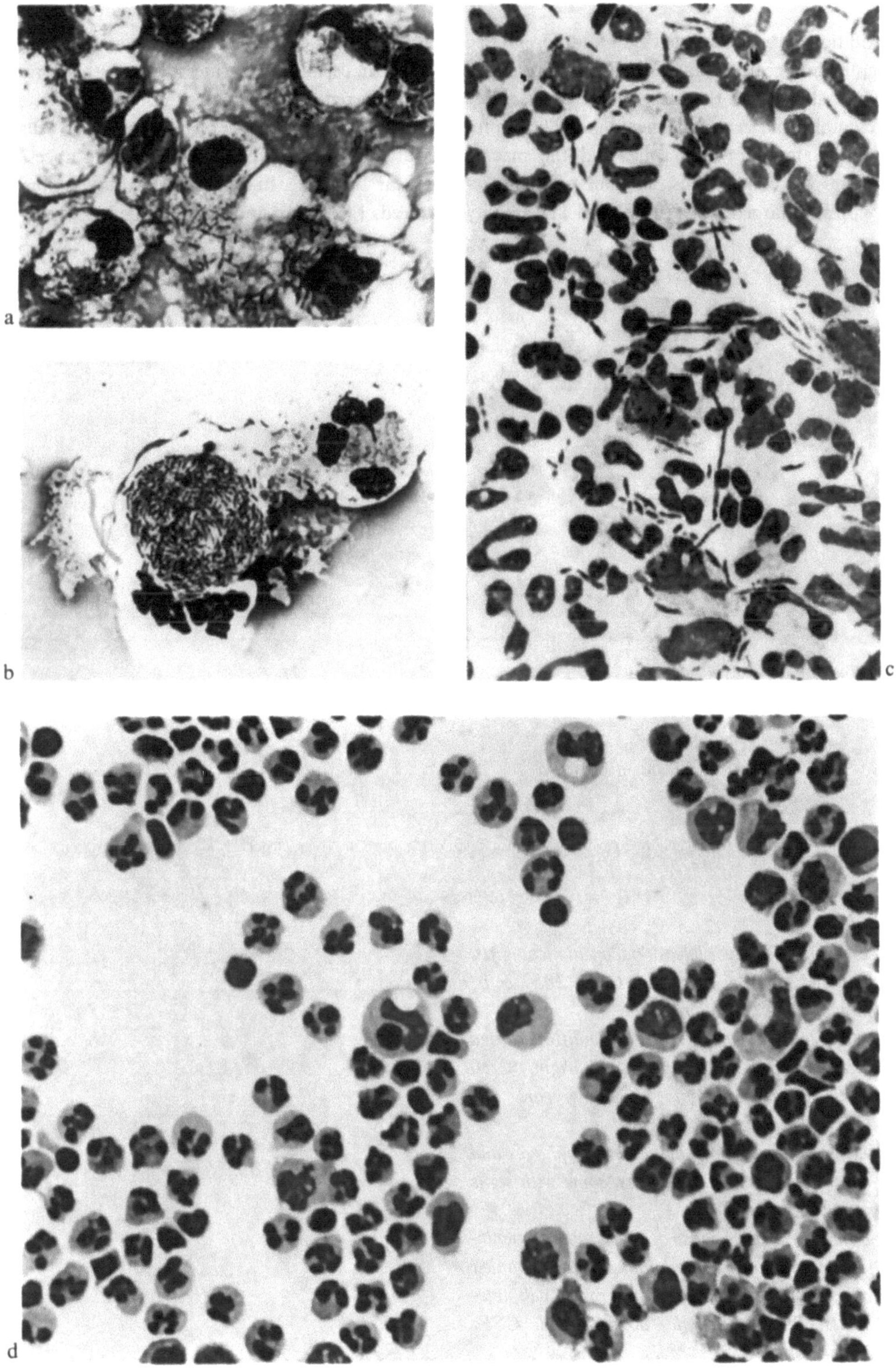
a
b
c
d

addition, large loose clusters of undifferentiated leptomeningeal cells are sometimes present.

The cellularity in the case of a *cerebral abcess* resembles that in chronic meningitis. The granulocytes are more numerous, and the macrophages are conspicuous, as some of them develop an enlarged signet ring shape. The activity of the lymphocytes is rather limited. Although there are certain similarities between this form and tuberculous meningitis, differentiation is sometimes possible. The total number of cells is usually greater when the site of the abcess is near the CSF. In general, the total seldom exceeds 1,000/mm^3 and is often under 300/mm^3.

Plate 16

a Early proliferative phase of bacterial meningitis. Patient: male, 4 years old. CSF: 2450 cells/mm^3. 95 mg% protein.
Many granulocytes. Nest of undifferentiated cells, probably of leptomeningeal origin. $\times 250$

b Brain abcess. Patient: male, 34 years old. CSF: 280 cells/mm^3.
Large number of monocytes, macrophages and neutrophils. Few lymphocytes, some with signs of transformation.

c Late proliferative phase of pseudomonas meningitis. Granulocytes predominate, accompanied by many lymphocytic-plasmocytic cell elements. Patient: female, 56 years old. CSF: 230 cells/mm^3.

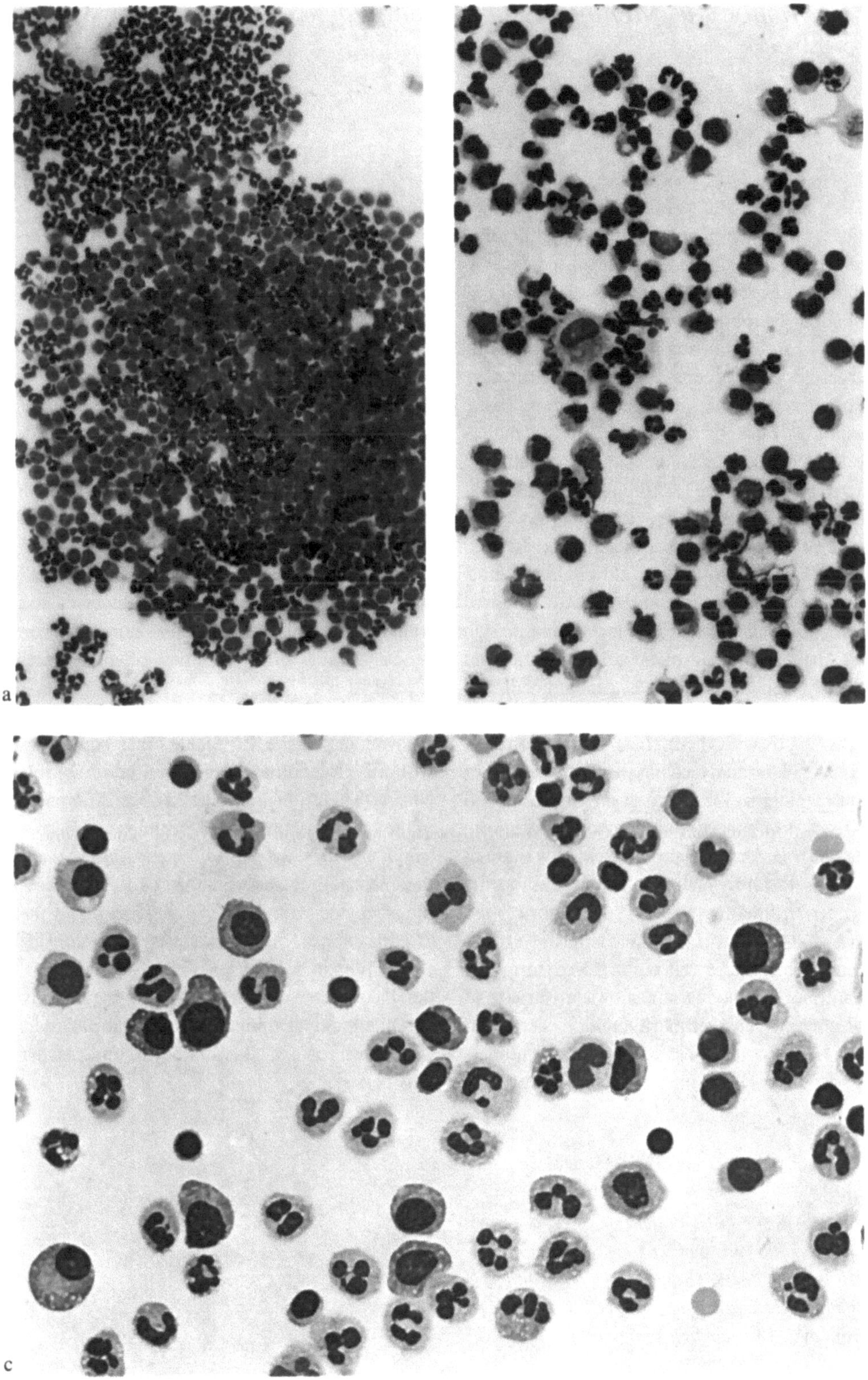

a b c

7. *Lymphocytic Meningitis and Meningoencephalitis*

In cases in which lymphocytic meningitis takes the usual, relatively benign course, the cause is usually a member of the enterovirus group.

A benign lymphocytic meningitis may also be observed accompanying mumps and almost all herpes zoster infections. Pleocytosis of the CSF is often very limited in cases of herpes zoster meningitis ($40/mm^3$), extensive in the majority of Coxsackie infections (up to $3{,}000/mm^3$) and does not exceed $1{,}000/mm^3$ in most forms of viral meningitis. There seems to be no correlation between the number of cells in the spinal fluid and severity of illness, although it is most often the meningitis accompanied by a limited pleocytosis that is clinically asymptomatic. On the other hand, mild pleocytosis is sometimes an early sign of a serious encephalitic illness. As a rule, one encounters the lymphocytic stage of meningitis with the first puncture. However, if the CSF is analyzed in the initial stage of viral meningitis, the observer may be surprised to find a pronounced granulocytosis in addition to a few transformed lymphocytes. This finding, common to all incipient infections, disappears abruptly, sometimes after 1 day, more often somewhat later, to be replaced by the typical lymphocytosis. The cell count can rise in the meantime. The cytogram shows a variety of transformed cells, primarily middle-sized and large lymphocytes. Plasma cells are also numerous, some polynuclear, others in the process of mitosis. A few dark-cytoplasmic and light-cytoplasmic lymphoid cell forms are conspicuous. Granulocytes, including a few eosinophils, account for a negligible fraction of the total number of cells. The final stages of meningitis are accompanied by a steady decline in the number of transformed lymphocytes and an increase in the number of degenerated macrophages. Finally, the specimen reveals an increased number of small lymphocytes, which are at best notable for their narrow cytoplasmic borders. Relapses, which sometimes occur in cases of viral meningitis, are always characterized by the renewed appearance of granulocytes and transformed lymphocytes. Final normalization of cell count and cellularity often requires several weeks.

One has to distinguish between benign viral involvement of the CNS and other conditions with an encephalitic, sometimes less favorable course. There are no characteristic differences in the cellularity of the CSF. Mumps and herpes simplex viruses, as well as the arborviruses, are the chief pathogenic organisms. In acute cases the pleocytosis may exceed $1{,}000/mm^3$, and the CSF may appear bloody as a result of the considerable admixture of red blood cells. In general, the pleocytosis declines with an increasingly encephalitic development of the viral disorder, although transformed lymphocytes and plasma cells persist.

Plate 17
Coxsackie meningitis. Patient: female, 25 years old. CSF: 700 cells/mm³. 65 mg% protein. Fifth day of illness. Granulocytes have almost completely disappeared. Extensive lymphocytic-plasmocytic cell reaction some mitoses.

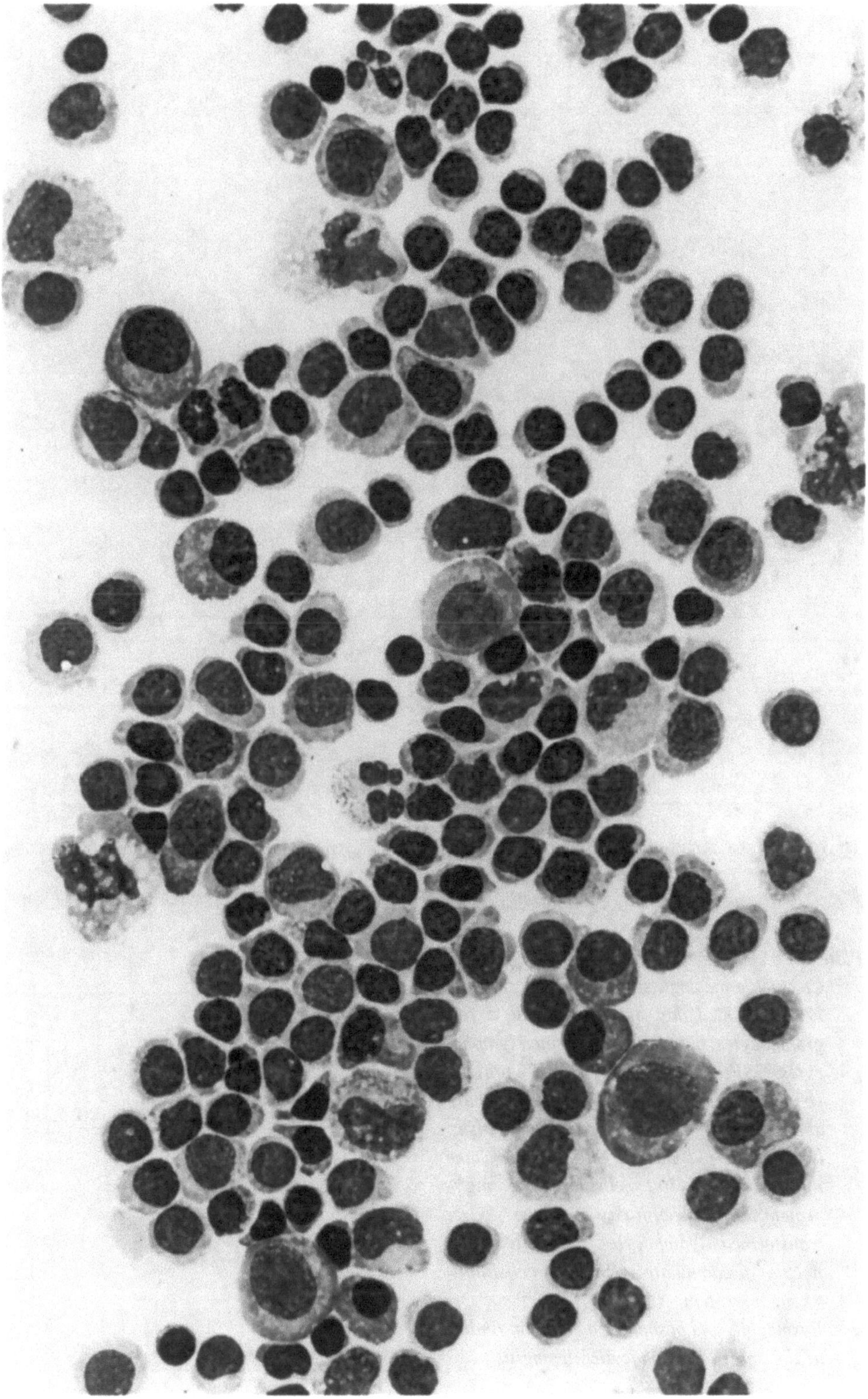

Plate 18

a–c *Coxsackie meningitis. Same patient as plate 17. (a) Third day of illness. Still a few granulocytes, but many transformed lymphocytes. CSF: 220 cells/mm³, 42 mg% protein. (b) Fifth day of illness. Granulocytes have almost completely disappeared. Extensive lymphocytic-plasmocytic cell reaction. Some mitoses. CSF: 700 cells/mm³, 65 mg% protein. (c) Fifteenth day of illness. Fewer transformed lymphocytes. Activated monocytes predominate. CSF: 40 cells/mm³, 45 mg% protein.*

d, e *Eosinophils and basophils, a common finding at the onset of lymphocytic meningitis.*

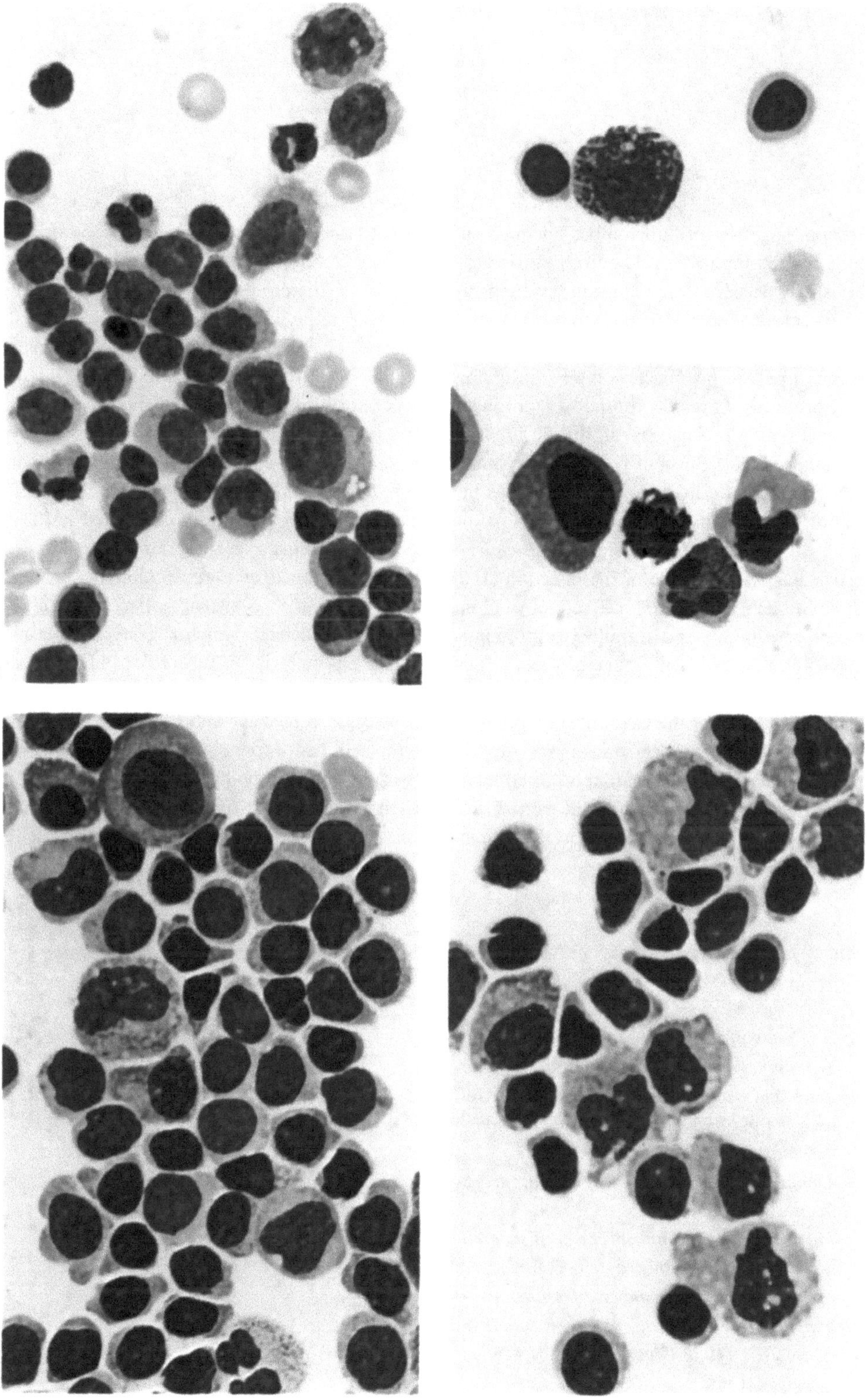

a e d b c

8. *Tuberculous Meningoencephalitis*

Prompt diagnosis of tuberculous meningoencephalitis and application of appropriate therapy are usually decisive in the clinical development of this severe illness. All diagnostically useful parameters have to be investigated at the onset of the illness, since bacteriologic evaluation often produces its results too late or produces no results at all. Analysis of the differential cell count of the CSF is of considerable value. Although there are no unequivocal findings indicating tuberculous meningitis, certain cytologic features reappear consistently and can be regarded as characteristic of the infection [140, 165, 176, 222]. The total cell count seldom exceeds 1,000/mm^3, and granulocytes predominate in the first exudative stage of tuberculous meningitis, as is also the case in many other infectious diseases of the CNS. The preponderance of granulocytes decreases more rapidly in tuberculous meningitis than in bacterial meningitis, but more slowly than in the viral form. Since the illness develops unobtrusively, admission to the clinic usually occurs several days after the development of the infection. Granulocytosis is, therefore, no longer evident at the moment of the first CSF puncture. Round cells assume greater importance, although many granulocytes are still present and remain in evidence until the end of the infection. A renewed outbreak of the inflammation is almost always accompanied by a sharp increase in the proportion of granulocytes. Such outbreaks can signify the affliction of previously unaffected parts of the leptomeninges or the brain, and they can also be evidence of a tuberculotoxic process. In any case, the coexistence of many granulocytes, transformed lymphocytes, and plasma cells is a peculiarity of tuberculous meningoencephalitis.

The round cells have more or less basophilic cytoplasm, an expression of their activity in resistant and allergic-hyperergic processes. It is often impossible to discriminate between various transformed lymphocytes and plasma cells.

Plate 19

Tuberculous meningitis. Patient: male, 16 years old.

- a, b *Sixth day of illness. No therapy as yet. Approximately equal proportions of granulocytes and transformed lymphocytes, as well as some plasma cells. CSF: 320 cells/mm^3, 120 mg% protein.*
- c *Lymphoid cell surrounded by various transformed lymphocytes.*
- d *Thirty days after commencement of tuberculostatic therapy. Only a few granulocytes remain. Declining number of transformed lymphocytes. Signs of monocyte mobilisation. CSF: 110 cells/mm^3, 90 mg% protein.*

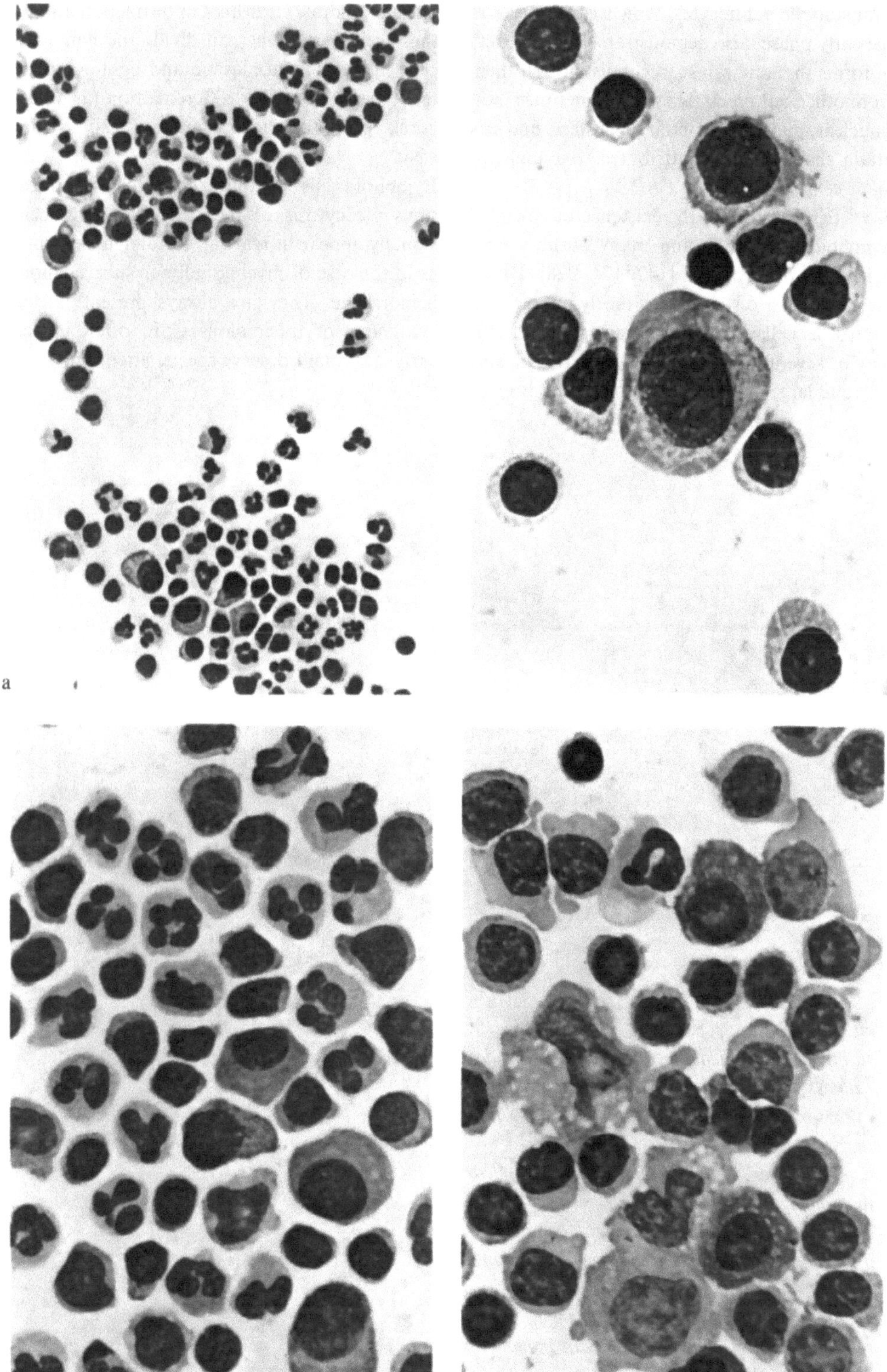

a c b d

Transformed monocytes, which can appear in the early phase, also demonstrate such a variety of forms that one can scarcely distinguish them from other cell types. They are sometimes multinuclear, may appear in large clusters, and can attain shapes that are attributed to Langhans' giant cells.

SAYK first referred to the existence of so-called lymphoid cells [165], and many authors confirmed his observations [140, 174, 228]. Its remarkable size distinguishes the lymphoid cell from other cells of similar character; the nucleus may be several times larger than that of plasma cells and large lymphocytes. However, it is often difficult to draw clear lines of distinction among these cell types. One can divide the lymphoid cells into dark-cytoplasmic and light-cytoplasmic types, but further differentiation (as Wieczorek proposed, for example [228]) is not easy.

Lymphoid cells can appear in all forms of infectious pleocytosis of the CSF [69] and occasionally appear in reaction to foreign elements, as in the case of myelography or subarachnoid hemorrhage. They are always present in the cytograms of tuberculous CSF, often in the early stage, and deserve special attention.

Plate 20

Fatal tuberculous meningoencephalitis. Third day of illness. No therapy as yet. Patient: male, 22 years old. CSF: 430 cells/mm³, 95 mg% protein.

- *a Large cluster of predominantly multinuclear macrophages.*
- *b Many transformed monocytes in loose clusters as well as some granulocytes. Phagocytized material in the cytoplasm.*
- *c Fluorescent stain with acridine orange. Strong red fluorescence indicates activity of monocytic and lymphocytic cytoplasm.*

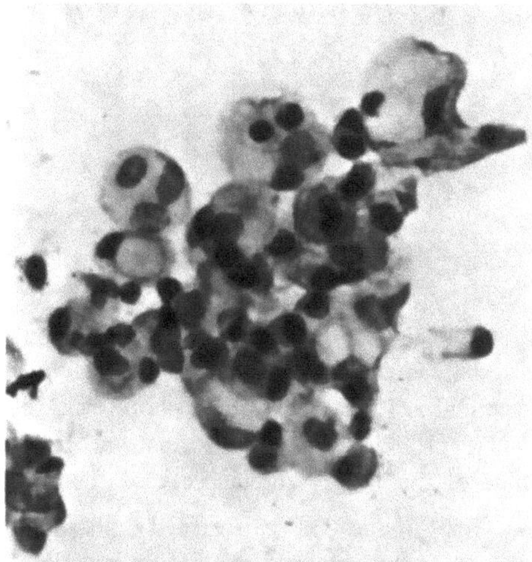

a

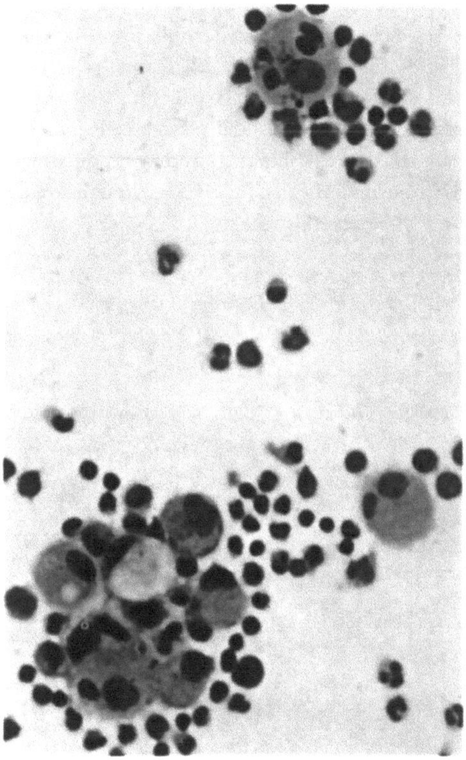

b

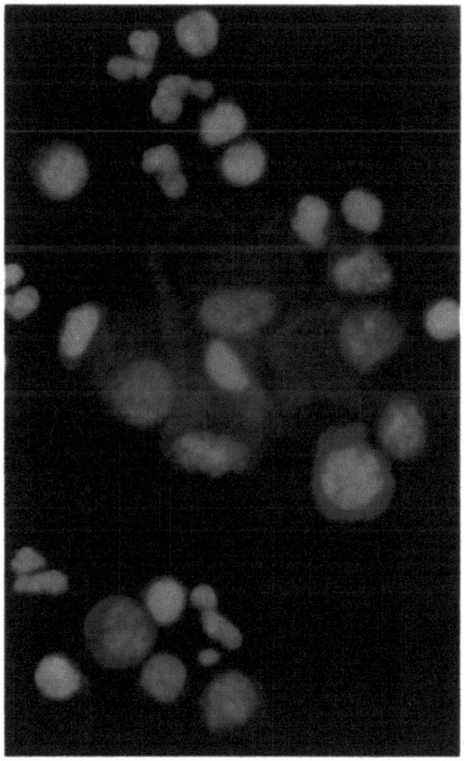

c

9. *Cerebral and Spinal Syphilis*

Syphilitic meningoencephalitis, tabes dorsalis, and progressive paralysis are not responsible for specific and characteristic cytologic findings in the CSF. However, cytologic monitoring of the CSF provides valuable information on the course of the illness and success of therapy.

Pleocytosis of the CSF can exceed 3,000/mm^3 in rare cases of syphilitic meningoencephalitis, but values under 500/mm^3 are more usual. As in the case of other infectious disorders, granulocytosis is characteristic of the initial phase. The granulocytes are soon replaced by transformed lymphocytes, plasma cells, and monocytes, though a limited number of granulocytes can usually be found over a long period of time. The granulocytes never attain the same proportions as in tuberculous meningoencephalitis. Penicillin therapy can quickly normalize the CSF.

The cell count usually does not exceed 40/mm^3 and may occasionally remain normal in cases of tabes dorsalis and progressive paralysis. Pleocytosis is usually greater in progressive paralysis than in tabes dorsalis. The cell count is usually a reliable indication of the severity of these diseases. Some authors have emphasized the small size of the lymphocytic cell forms, which are almost always in evidence [83, 133, 138]. A few monocytic cell forms and plasma cells [64, 173] are present. Penicillin therapy may produce cytologic normalisation of the CSF, though usually only after several weeks.

Plate 21

a, b Syphilitic meningoencephalitis. About 2 weeks after first signs of illness. Lymphocytes predominate. Many cells with cytoplasmic transformation. Patient: female, 34 years old. CSF: 180 cells/mm^3, 65 mg% protein.

c, d Syphilitic meningoencephalitis. Same patient as Fig. a and b, 3 days after initiation of penicillin therapy. Still many transformed lymphocytes. Plasmocytic cells. Signs of monocyte mobilisation. CSF: 90 cells/mm^3.

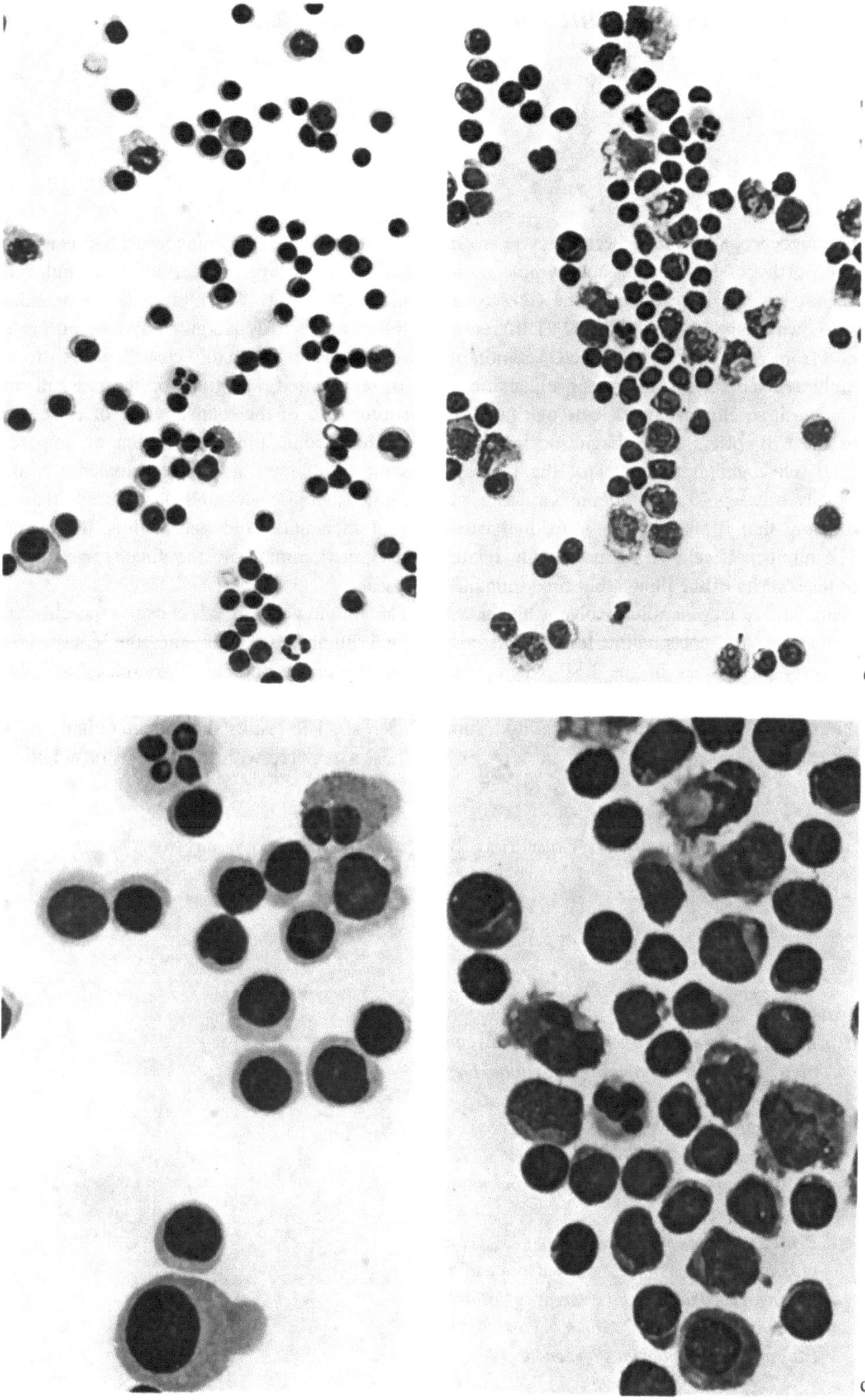

a c b d

10. Meningoencephalitis in Behçet's Disease

Meningoencephalitis in Behçet's disease is rare but worthy of mention as an example of an allergic-hyperergic reaction of the CNS to an unknown agent [3, 4, 15, 200, 232]. Differentiation from other infections of the CNS, notably multiple sclerosis, is sometimes problematic.

Under these circumstances, cytologic analysis of the CSF offers some diagnostic help. The total cell count at the onset of the illness is usually between 30 and 150/mm^3 and can rise to more than 1,500/mm^3 in individual cases. The number of cells is not necessarily related to the severity of the illness, but predominantly meningitic cases generally involve a higher cell count than the encephalitic forms. One also observes pleocytosis of the CSF when additional symptoms are other than neurologic in nature, as is the case in erythema nodosum, for example.

The acute stage of meningoencephalitis is associated with a large proportion of granulocytes in the CSF (up to 90%), but within 1 or 2 days the granulocytosis can give way to a variegated cell picture in which all hematologic cell forms are represented. The granulocytes may still constitute 10% of the total number of cells, even in the chronic phase or during an apparent standstill of the illness. There are fewer transformed lymphocytes and plasma cells than in viral meningitis and encephalitis. Monocytic cells predominate in the final stage of the disease.

The findings described above—especially the total number of cells and the conspicuous increase in the number of granulocytes—allow a relatively certain differentiation from multiple sclerosis. The results show greater similarity to those associated with tuberculous meningitis.

Plate 22

a, b Behçet's disease. Patient: male, 36 years old. (a) Purely meningitic reaction. First sign of illness 16 days earlier. CSF: 1080 cells/mm^3. Neutrophils throughout. A few monocytes. (b) Six days later. CSF: 35 cells/mm^3. Mainly monocytes and small lymphocytes.

c, d Behçet's disease. Patient: male, 32 years old. CSF: 70 cells/mm^3. Second relapse of meningoencephalitis. First signs of illness 18 months previously. Granulocytes, many transformed lymphocytes and monocytes.

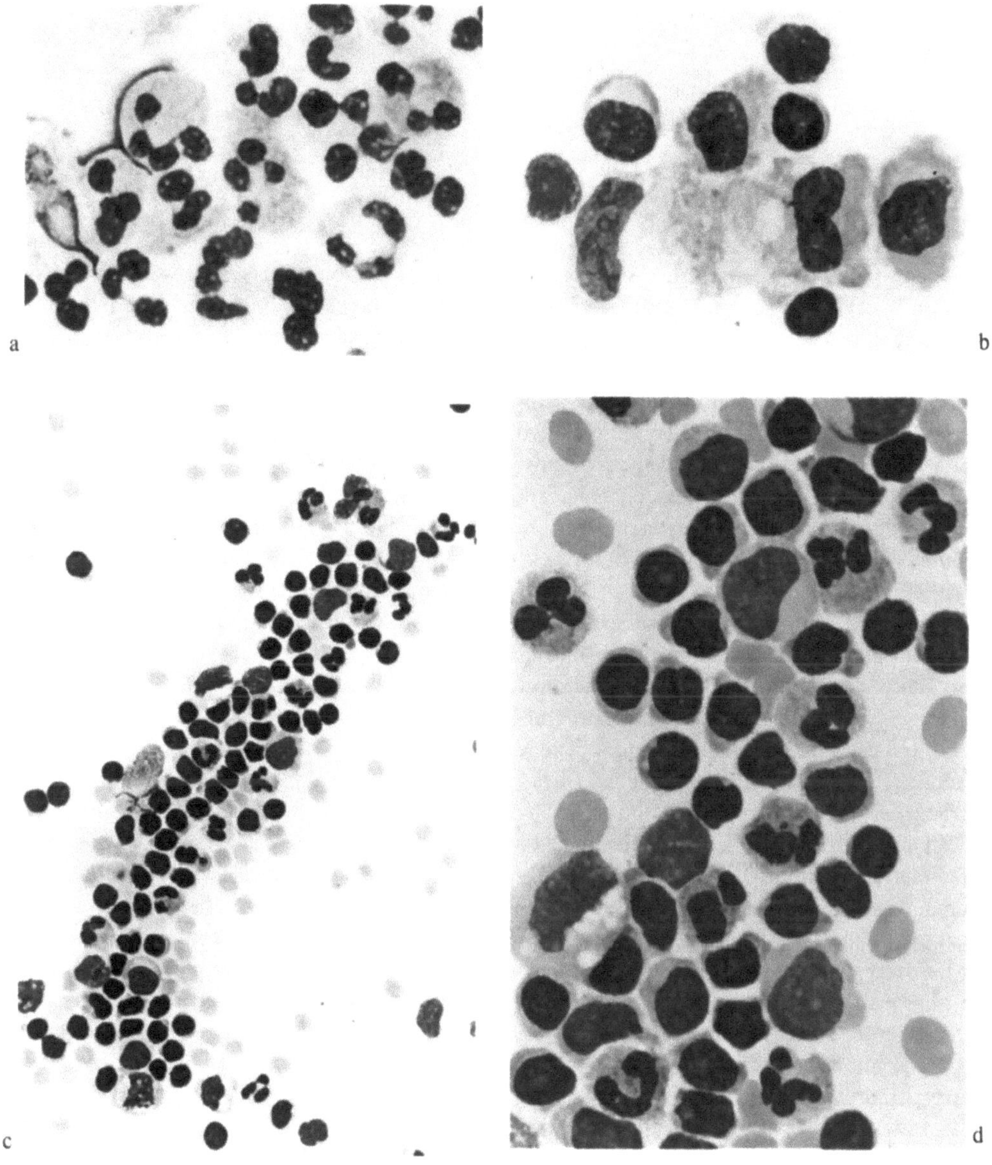
a
b
c
d

11. Giant Cells

The term "giant cells" was established on the basis of morphologic criteria. Because of their extraordinary size, these cells are inconsistent with the cellularity of normal CSF, and their appearance almost certainly indicates a pathologic process. One can distinguish between two sorts of giant cells according to their origins: those which stem from the tissue of malignant tumors, and those which have developed from leptomeningeal stem cells. Both are impressive for their enormous variety of form.

The existence of monocytic or histiocytic giant cells indicates chronic irritation of the meninges, which can have various origins. The most common cause is the introduction of a "foreign" element into the CSF. Examples include the presence of blood after subarachnoid hemorrhage, of cerebral tissue and blood following a cerebral contusion, of contrast medium, or of a ventricular drain. These cells usually contain several, sometimes very many, round or oval nuclei which lie individually or in groups in the large cytoplasmic body.

It is probable that the cells are multinuclear when they are released from leptomeningeal tissue. Amitotic cell division in the CSF is rarely a source of multinuclear cells. The cytoplasm itself stains a relatively weak blue-gray. The presence of certain vestiges, such as fat vacuoles, hemosiderin, and hematoidin, allows conclusions as to the origin and the phagocytic activity of the cells.

In addition to the giant cells, which may be regarded as special forms of macrophages, there are other types which appear in the course of subacute or chronic, specific or unspecific infections of the CNS [127, 164, 225]. These are remarkable for a more pronounced cytoplasmic coloration, to the point of extreme cytoplasmic basophilia. A high rate of mitosis is an additional indication of the activity of these cells. They can occasionally assume forms reminiscent of megakaryocytes. Giant cells with strong cytoplasmic basophilia are to be found in cerebral mycosis [97] and, occasionally, in cases of encephalitis in children. The latter are probably caused by viral infections. Langhans' giant cells also appear rarely in tuberculous meningoencephalitis, but they probably have little prognostic significance. The cells can be extremely large, have uniformly large, well-contoured, excentrically-located nuclei, and loose cytoplasm.

Plate 23

Giant cells.

- *a Hydrocephalus of unknown etiology. Patient: male, 4 years old. CSF: 15 cells/mm^3. Giant cell with many horseshoe-shaped nuclei. Frayed cytoplasmic border.*
- *b Cerebral trauma at delivery. Hydrocephalus. Patient: male, 8 weeks old. CSF: 40 cells/mm^3. Undifferentiated "reticular" giant cell with 2 isomorphic nuclei, rich in chromatin. Basophilic cytoplasm.*
- *c Cerebral contusion. Patient: female, 18 months old. CSF: 90 cells/mm^3. Probably cell of the choroid plexus.*
- *d Hydrocephalus. Patient: male, 3 years old. 2 Langhans' giant cells abutting a third multinuclear giant cell.*
- *e Condition 50 days after Pantopaque myelography. Patient: female, 62 years old. CSF: 24 cells/mm^3, 74 mg% protein.*
- *f Chronic meningoencephalitis. Patient: female, 56 years old. CSF: 45 cells/mm^3, 120 mg% protein.*

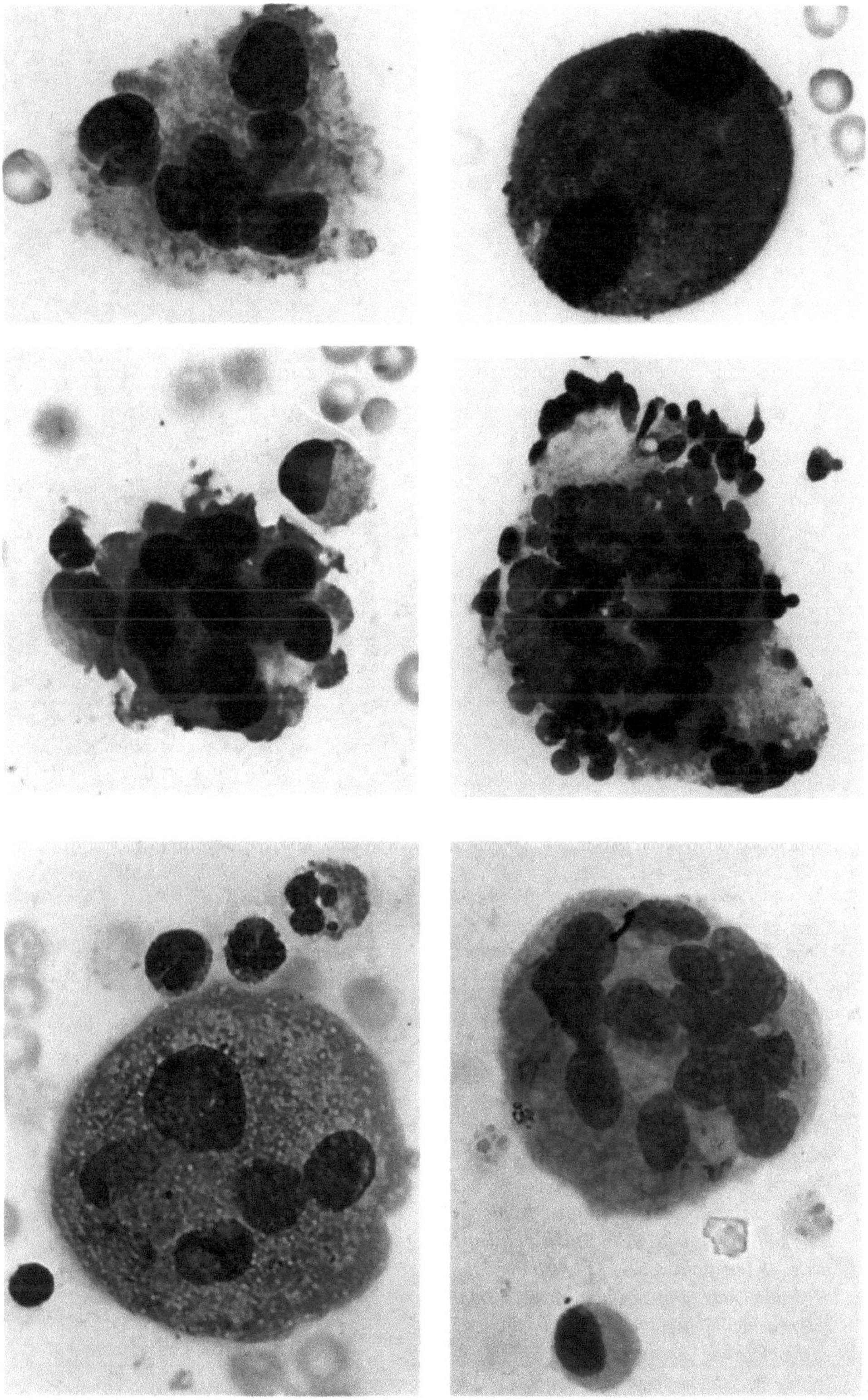

a b c d e f

Plate 24

Giant cells.

a, b *Recovery phase of meningoencephalitis. Many giant cells, with cytoplasmic basophilia and conspicuously high rate of mitosis.*

c *Tuberculous meningoencephalitis. Patient: male, 22 years old. CSF: 430 cells/mm^3. 96 mg% protein.*

d *Chronic meningoencephalitis after frontobasal skull injury and CSF fistula. Patient: male, 48 years old. CSF: 32 cells/mm^3.*

e *Hydrocephalus. Spitz-Holter drain. Giant cell resembling megakaryocyte among many eosinophils.*

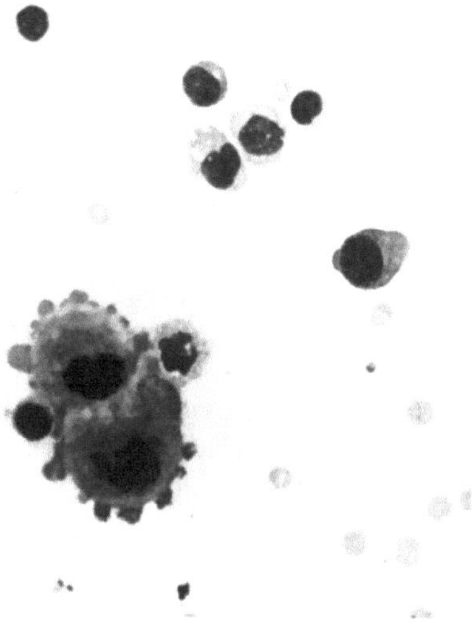

a

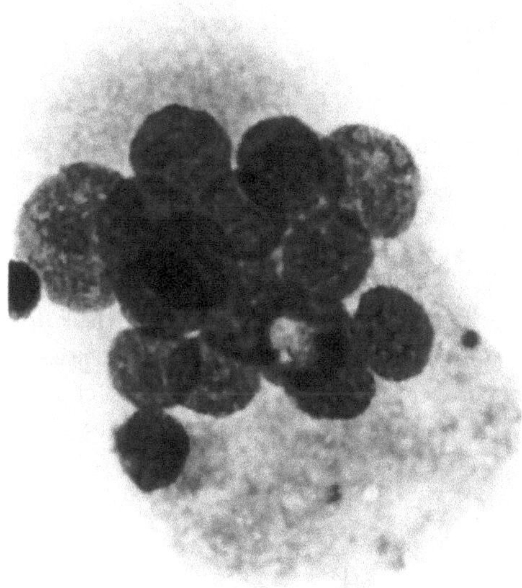

d

b

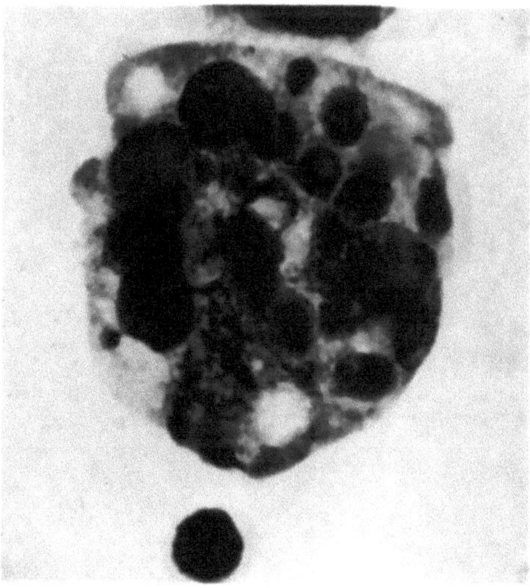

c

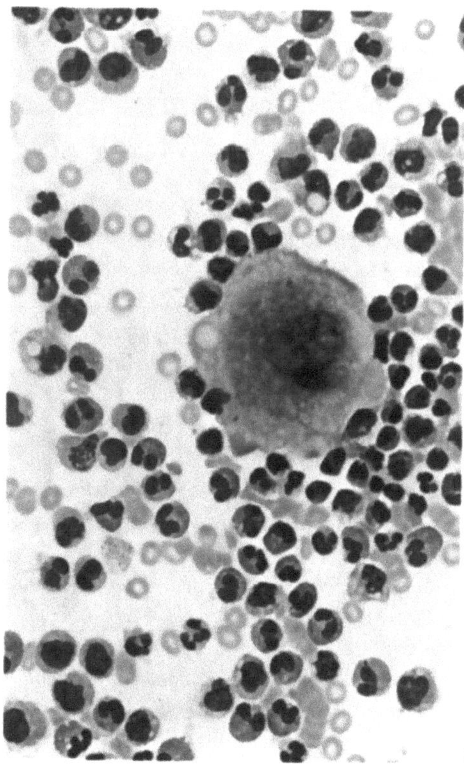

e

12. Tumor Giant Cells

It is sometimes difficult to differentiate between giant cells which stem from malignant tumors and those associated with infection. The giant cells that appear in cases of meningoencephalitis in children show great mitotic activity and are most often mistaken for cells of tumorous origin. However, tumor cells usually show other signs of malignancy—cell clusters and enlarged nucleoli, for example—in addition to enormous size. The giant cells in glioblastoma multiforme [19] and in numerous carcinomas are most conspicuous, and they appear almost without exception in adenocarcinoma. In the latter case, the cells are sometimes greatly enlarged signet-ring cells. Giant cells may also appear in other tumors, such as ependymoma and plexus papilloma.

Plate 25
Tumor giant cells.

- a *Polymorphic pinealoma. Note the bean-shaped to horseshoe-shaped nuclei. Patient: female, 28 years old. CSF: 25 cells/mm^3.*
- b *Ependymoma of the fourth ventricle. Previous radiation therapy. Patient: female, 8 years old. CSF: 15 cells/mm^3.*
- c, d *Malignant, polymorphic choroid plexus papilloma. Patient: female, 65 years old. CSF: 14 cells/mm^3. (c) May-Grünwald-Giemsa stain. (d) Fluorescent stain with acridine orange.*
- e *Cerebral metastases of a melanoma. Patient: female, 56 years old. CSF: 15 cells/mm^3.*
- f *Adenocarcinoma of the stomach. Patient: male, 33 years old. CSF: 110 cells/mm^3.*

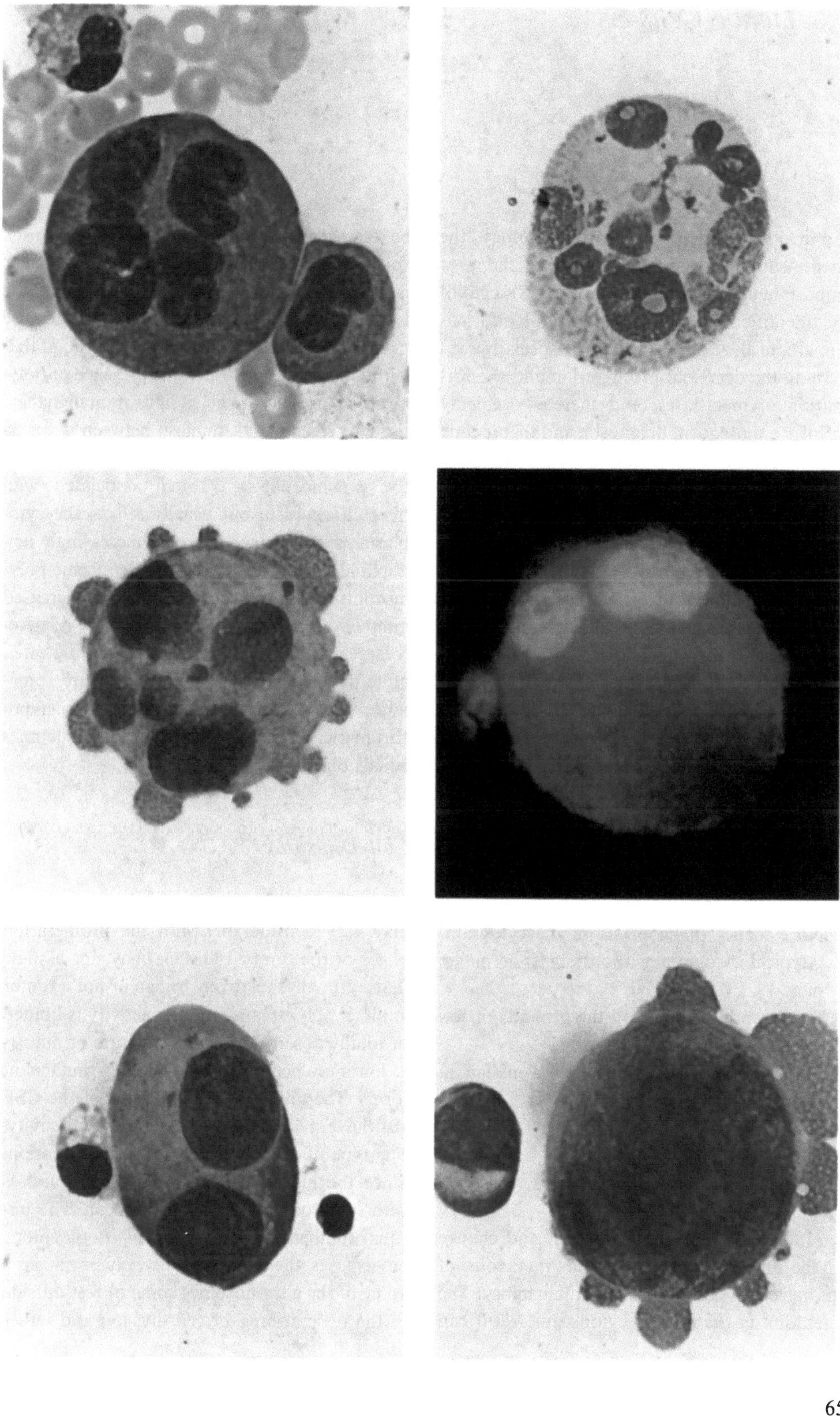

a b c d e f

13. Tumor Cells

As in other fields of exfoliative cytology, the identification of tumor cells is one of the most important and difficult activities in CSF cytology. Identification can provide useful information about the suspected tumor, especially concerning the degree of proliferation and the formation of metastases, and thereby decidedly influence subsequent diagnostic and therapeutic procedures. In some cases, CSF cytology can aid in the early recognition of malignant processes, especially in leukoses. In other cases, it can confirm the clinician's suspicions, and, in still others, it can give an unexpected new direction to clinical investigations.

The diagnosis of tumor cells is founded on known, generally valid criteria of malignancy which should consider the individual cell, cell groups, and the individual cell in comparison with other cells of the same substrate. The reliability of the diagnosis is proportional to the number of cells investigated and the number of malignant signs found. The following classification for cytologic findings, a modification of PAPANICOLAOU'S system [145], has proved useful:

1. No evidence of neoplasia in the cytogram.
2. Atypical cytogram without signs of malignancy.
3. Suspicion of a tumor on the grounds of few signs of malignancy.
4. No doubt as to the existence of a malignant neoplastic process.

Nucleus

An increase in the size of the cell and change in the nuclear-cytoplasmic ratio in favour of the nucleus are both criteria of malignancy. The alteration in the nuclear-cytoplasmic ratio can be so great that some tumor cells seem to consist solely of bare nuclei. In addition, the nucleus is conspicuous for its hyperchromatism, and the loose, often inhomogeneous chromatic texture may vary considerably from one cell to another within the same substrate. The presence of polymorphic nuclei can be a sign of malignancy, though one has to distinguish between degenerative and neoplastic polymorphism. Degenerative polymorphism is often associated with hyperchromatism but usually affects the cytoplasm as well, which becomes increasingly acidophilic. This is not the case in neoplastic polymorphism. The nucleus contains an increased number of nucleoli, which may occupy relatively large areas of the karyoplasm. They are often quite distinct from the nucleus and are sometimes conspicuously hyperchromatic or hypochromatic. Their polymorphism sometimes equals that of the nuclei.

Cell Division

The rate of mitosis is increased, but this finding may vary considerably with the proliferation phase of the tumor cell. One may not assume that a growth is relatively benign or not a tumor at all, simply because mitotic activity is limited or totally absent. Generally, the rate of mitosis is lower in benign tumors than in malignant forms. The unfavorable conditions of the CSF also have a limiting effect on mitotic activity. The type of mitosis deserves special attention. Since the nuclei are often polyploid, mitoses with three or more poles may occur. The unequal number of wandering chromosomes alone predisposes the new cells to polymorphism. Some of the chromosomes come to rest outside of the normal area of cell division and either degenerate or appear later as separate nuclear fragments. Other chromosomes appear suspiciously swollen, clotted, or drawn out to long filaments. Amitotic cell division occurs but seems to be rare. Nuclear divisions, amitotic or mitotic, also occur without cytoplasmic division, so that polynuclear giant cells result.

Cytoplasm

The cytoplasm stains dark blue with panoptic preparation. The luminous red with acridine orange indicates an increased RNA content and increased cytoplasmic activity. The border of the cytoplasm can be smooth, sharply indented, or even frayed in appearance, while the cytoplasm itself can be limited to a narrow margin as a result of the relatively large size of the nucleus (see above). On the other hand, some tumor cells, especially those of epithelial origin, are remarkable for a large, differentiated cytoplasm and a relatively small nucleus. Cytoplasmic inclusions, up to entire signet-ring cells, are more typical of epithelial provenance, whereas homogeneously stained, relatively undifferentiated cytoplasm indicates mesodermal origin. More specific conclusions on the origins of the cells are usually beyond the capabilities of cytologic investigation. Finally, signs of phagocytosis are evidence against malignancy.

All of the criteria of malignancy listed above are diagnostic indicators but are not to be regarded as absolutes. It may be difficult to find them in some cases and impossible in others. One tumor cell may present all the criteria, while another may be concealed behind an apparently harmless facade. Both transformed monocytes and lymphocytes are capable of exhibiting an extraordinary variety of shapes, extremely large size, multiple nuclei, nuclear and cytoplasmic basophilia, large nucleoli, and atypical mitoses [19]. One must, therefore, exercise great care in the choice and formulation of the diagnosis. There is no unequivocal criterion of malignancy for the individual cell; the actual criterion—invasive and destructive growth—can only be recognized in a tissue section. Although an experienced cytologist can frequently make a diagnosis with considerable certainty, it is just as often the case that he can only report his suspicions.

One can employ additional cytologic analyses, as well as panoptic staining, in order to establish new parameters of malignancy. The PAS reaction has proved valuable in demonstrating the presence of abnormal amounts of mucopolysaccharides, especially in the cytoplasm of adenocarcinomas [142]. Cresyl blue stain has special affinity for nucleic acids and can confirm suspicions of the existence of a tumor.
Fluorescent staining with acridine orange and subsequent critical evaluation under ultraviolet light can provide valuable insights into cytoplasmic activity [18, 44, 181]. Acridine orange displays a fluorescent metachromasia in which low pigment concentrations appear green and high concentrations red. Since acridine orange has greater affinity for the RNA of the cytoplasm than for the DNA of the nucleus, the cytoplasm appears red and the nucleus green to yellow-green under ultraviolet light. The DNA content of a cell type is a constant, but the RNA content varies with cytoplasmic activity, so that the nuclei of a cell population appear a uniform green to yellow-green, while the cytoplasm may vary from a dull redbrown to luminous red. One can recognize the cytoplasm of malignant cells, whose RNA content is presumably high, by the luminous red color. The nucleoli, which are often numerous and enlarged, contain RNA and appear luminous as well. These characteristics are not unique to tumor cells; other cells which proliferate rapidly under certain conditions and which have a high RNA content (plasma cells, lymphocytes and transformed monocytes, for example) also give an intense red fluorescence. This tends to diminish the diagnostic value of acridine orange in tumor cytology, but the method can contribute a great deal to cytodiagnosis if the results are correctly interpreted.
Autoradiography has also been successfully introduced in the cytologic diagnosis of tumors.

Tritiated thymidine has proved to be the most suitable tracer, and the cell turnover rate can be determined by incubating a cell population with tritiated thymidine in vitro [28]. The substance has a strong affinity for DNA in the cell and is incorporated into cellular structures which are capable of proliferation and actively involved in DNA synthesis. The result is a direct proportionality between the amount of tracer substance and the cell turnover rate. A positive result is not specific evidence of a tumor, but, in our experience, the method is useful in the identification of tumor cells [98].

Localization and degree of malignancy of primary and secondary cerebral growths determine whether malignant cells will be exfoliated into the CSF. As is to be expected, one can more often find tumor cells when the growth is located close to the CSF, but more distant localization of the tumor does not exclude the possibility of finding specific cells in the preparation [201]. The histologic characteristics of the tumor are decisive as well as its tendency to grow and infiltrate and its ability to form loose, easily dispersed cell agglomerations.

Many authors emphasize that the chances of demonstrating tumor cells are greater with ventricular or suboccipital CSF collection or with pneumencephalography than with lumbar puncture [14, 115, 118, 130]. This may be true, especially if one considers the unsuitable conditions endured by a tumor cell exfoliated in the ventricles or the cisternae before it reaches the lumbar area. On the other hand, the lumbar area is a collection basin for all cells of the CSF. The cells may even be concentrated there and collected with the first portion of the CSF.

13.1. Primary Tumors of the CNS

Authors differ on the question of how often tumor cells can be identified in the CSF in cases of primary brain tumor [124, 166, 226, 230]. As we have already emphasized, cytologic results depend on the location of the tumor and of the puncture and, of course, on the amount of CSF that is cytologically analyzed. It is likely that tumor cells can be identified in less than 15% of cases of primary brain tumor, if the CSF sample is collected by lumbar puncture. Some reports claim much better results [11, 20, 96, 130, 201, 221].

The limited tumor cell yield is not surprising, as the majority of primary brain tumors, examined against histologic criteria, are relatively benign neoplasias with limited mitotic activity and no tendency to produce metastases. They have only a very limited tendency to exfoliate cytologically differentiable cells into the CSF space. Finally, it is difficult to identify cells of benign tumors, which lack histologic characteristics of malignancy, in the CSF.

Some authors claim that the tumor cells of glioblastomas and medulloblastomas possess a few relatively constant characteristics which permit certain conclusions as to their origin, even on the part of the most conservative diagnostician [20, 22, 130, 193, 227]. In the following section of illustrations, we shall try to indicate certain characteristics of CSF cells from these and other brain tumors, but it must be emphasized that a definitive diagnosis is rarely possible, since the variety of cytologic phenomena is enormous.

Neurosurgical intervention and cytochemical or radiation therapy of a tumor or its metastases can change the mode of growth of the cells. The result is increased polymorphism and diminished clarity of the original cell characteristics.

13.1.1. Ependymoma

Ependymomas occur primarily in adolescents. Although one of the less common brain tumors, positive tumor cell findings in the CSF are not rare. This type of tumor apparently has a limited tendency to exfoliate cells into the CSF space. The chances of finding tumor cells increase if the tumor originates in the cerebrum, if the tumorous material has invaded the ventricles or the cisternae, or if the mode of proliferation has changed after an operation [202, 203]. Diffuse metastasis through the leptomeninges is not exceptional in some forms of this tumor. In such cases, tumor cells can regularly be found in the CSF, and their identification is simplified, since many criteria of malignancy are fulfilled.

Identification can be difficult if the cells derive from ependymoma, which do not proliferate very actively. Differentiation from cells of normal ependyma [35], from choroid plexus cells or from cells of the mononuclear phagocytic system may be especially problematic. The tumor cells are large and are often grouped in loose clusters, which sometimes resemble epithelium. The cytoplasm has a foamy appearance and stains an inhomogeneous, soft blue-gray. The nucleus can vary greatly: it is relatively small, often round, sometimes pyknotic, sometimes centrally-located, otherwise located excentrically or at the extreme border of the cell [22, 165, 195, 227]. Mitosis may appear.

In addition, one can find cells with a large nucleus, rich in chromatin, and with a narrow border of basophilic cytoplasm. They differ from lymphocytic and plasmocytic cell forms in their nuclear-cytoplasmic ratio.

Pleocytosis almost always accompanies tumor cells, and monocytes and macrophages predominate. They seem to exercise a defensive function.

Plate 26

a–c *Cerebral ependymoma. Patient: female, 20 years old. CSF: 40 cells/mm³. (a) and (b) Tumor cells with large inhomogeneous cytoplasm. Signet ring cells. Round, oval, or kidney-shaped nuclei. Hardly any nucleoli to be seen. (c) Lymphoid tumor cell (left border) with large nuclei, rich in chromatin, and a small, basophilic border of cytoplasm.*

d *Cerebral ependymoma after neurosurgical intervention. Patient: male, 17 years old. CSF: 12 cells/mm³.*
3 Tumor cells in a loose cluster. Conspicuously wide cytoplasmic border.

e–g *Cerebral ependymoma. Patient: male, 24 years old. CSF: 10 cells/mm³. Tumor cells of various types. Cell in (f) has the most pronounced epithelioid character.*

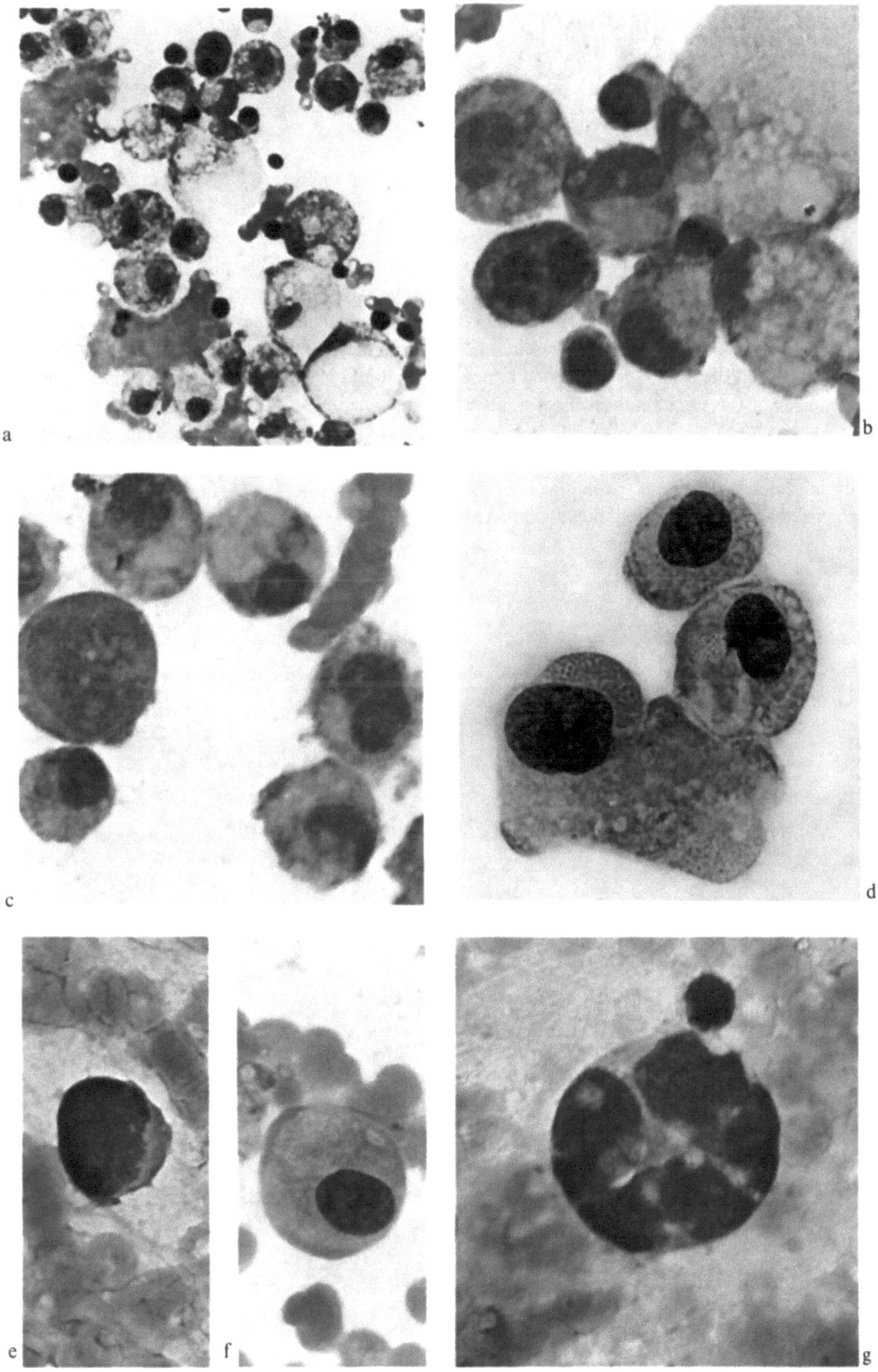
a
b
c
d
e
f
g

13.1.2. Choroid Plexus Papilloma

These rather rare tumors originate in the epithelium of the choroid plexus. It is possible to differentiate histologically the isomorphic benign papilloma from the polymorphic malignant form, which is comparable to a carcinoma. Cells or cell clusters may be exfoliated, since both forms of the tumor are in direct contact with the CSF. Metastases may develop after transport in the CSF. It is doubtful whether cells of the benign plexus papilloma can be recognized as tumor cells, though exfoliated cells of the malignant types are easily identified in the CSF [48]. The cells appear individually or in epithelioid clusters and occasionally display the original plexus structure [167, 224]. There may be similarities with ependymoma. The round-to-oval nuclei are rich in chromatin and sometimes display strongly contoured nucleoli. The cell body is large and has a coarsely-grained cytoplasm with a rather inconstant affinity for stain.

Plate 27
Polymorphic choroid plexus papilloma. Patient: female, 65 years old. CSF: 14 cells/mm³.
- *a An epithelioid cluster of isomorphic cells similar to those of ependyma or choroid plexus.*
- *b Organization of cell cluster less regular than in a.*
- *c Autoradiography after in vitro incubation with H^3-thymidine. Centrally located cell with distinctly marked nucleus. Focus in the plane containing silver granules.*
- *d Fluorescent staining with acridine orange. Strong red fluorescence in the cytoplasm of almost all the cells as evidence of increased cytoplasmic activity.*

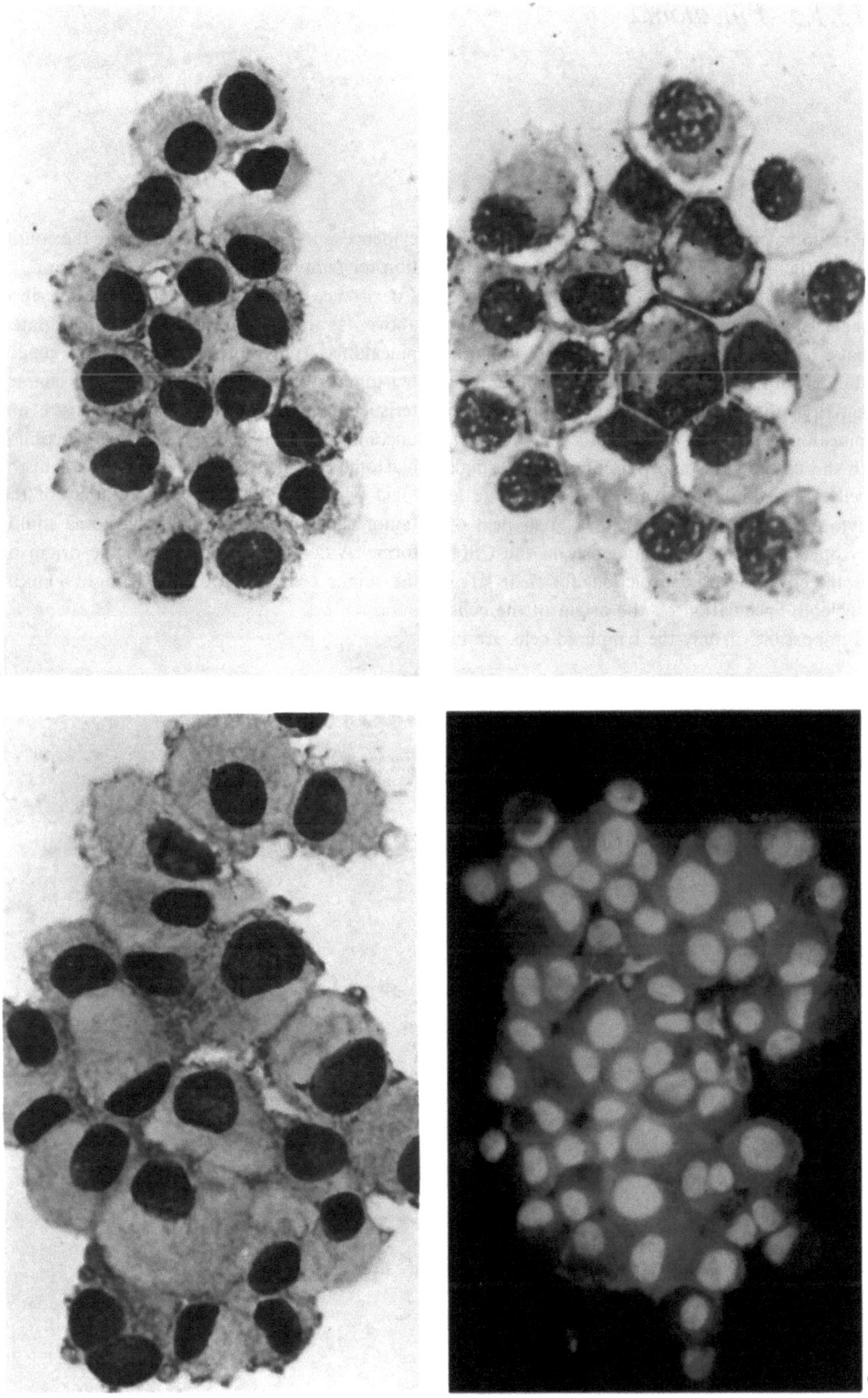

a c b c

13.1.3. Pinealoma

Pinealoma cells can exfolitate in the CSF without difficulty as a result of their location [10, 122, 137, 265]. Tear metastases [238] and diffuse metastasis through the leptomeninges [206, 219] have been described. It is more likely that the tumor forms with the greatest proliferative activity—the anisomorphic and polymorphic pinealomas—appear in the CSF.

In the case of anisomorphic pinealoma, small cell types, similar to lymphocytes, with little cytoplasm and large cells with a great deal of cytoplasm, appear simultaneously in the CSF. Both cell forms are conspicuous for their large nucleoli. Speculation on the origin of the cells is impossible, if only the lymphoid cells are in evidence, and it is possible to mistake the condition for lymphocytic meningitis or leukemia.

Coexistence of both tumor cell forms described above is uncommon in the undifferentiated pinealoma (polymorphic pinealoma, pineoblastoma). Wide variation in cell size is characteristic of this tumor, and giant cells are not uncommon. Similarity to cells of medulloblastoma has also been established with histologic evidence [238], while the cytology of the tumor can remind one of glioblastoma multiforme. A definitive judgment on the origin of the tumor cells is impossible without clinical data.

Plate 28
Anisomorphic pinealoma. Patient: male, 27 years old.

a, b *CSF: 215 cells/mm^3. Many isomorphic cells, similar to lymphocytes. Conspicuously large nucleoli.*

c, d *CSF: 400 cells/mm^3. 12 mg% sugar. Puncture 5 days later. Obviously disturbed cellularity. Reappearance of lymphoid tumor cells. A few large, hyperchromatic cells, some polynuclear. Phagocytes.*

e *Loose tumor cell group. Prominent nucleoli.*

f *Fluorescent staining with acridine orange. The large cell types demonstrate especially strong red fluorescence in the cytoplasm.*

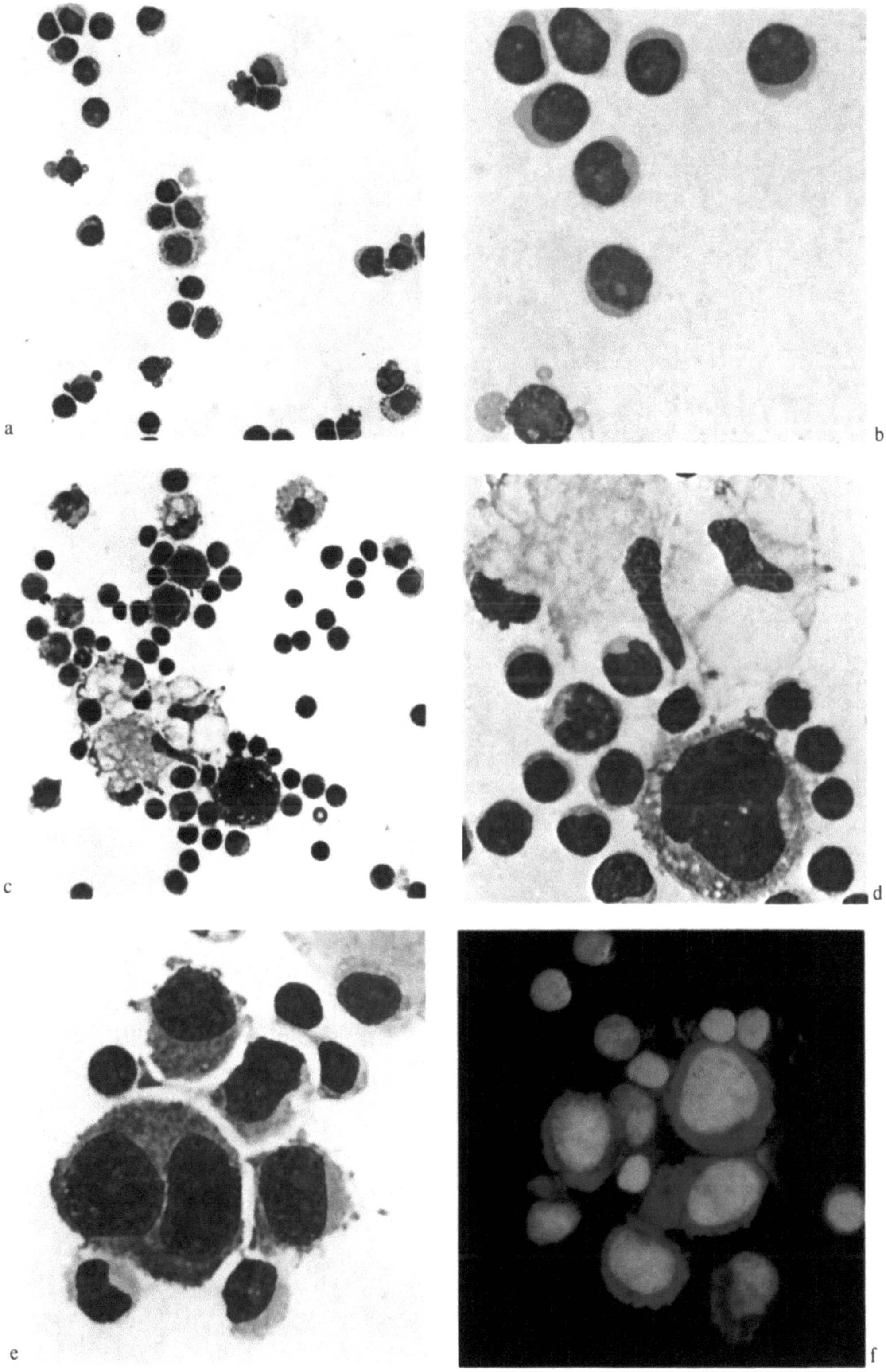
a
b
c
d
e
f

Plate 29

Polymorphic pinealoma. Patient: female, 28 years old. CSF: 25 cells/mm^3.

a Cluster of polymorphic tumor cells.

b Two small tumor cells with considerable shift in nuclear-cytoplasmic ratio.

c Tumor giant cell with many nuclei and nuclear buds, some of them bizarrely shaped.

d Loose tumor cell group.

e Tumor cell cluster. For the most part confluent cytoplasmic borders. Prominent nuclear polymorphism. Similarity to medulloblastoma in nucleus and cytoplasmic texture.

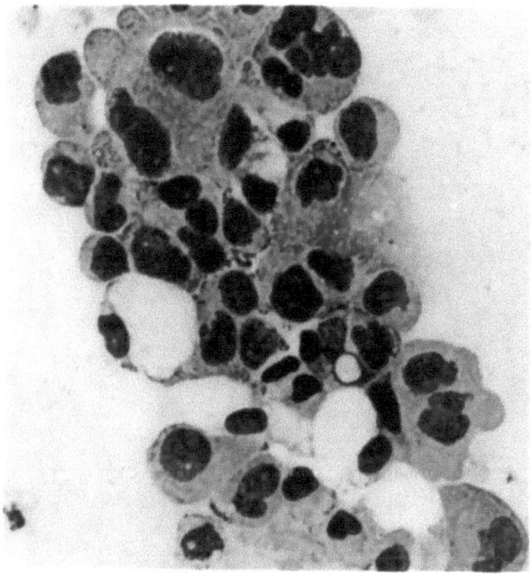
a

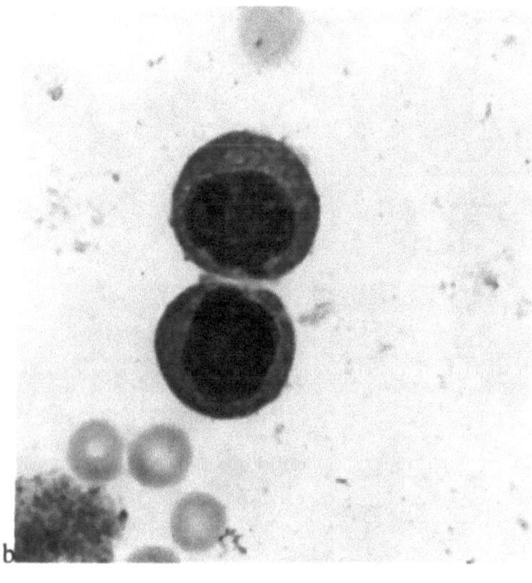
b

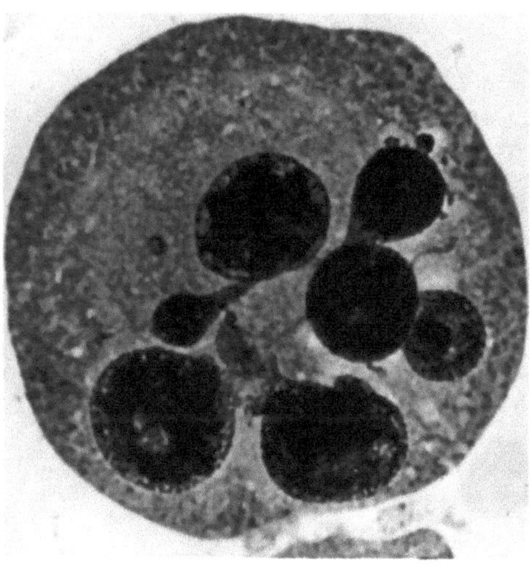
c

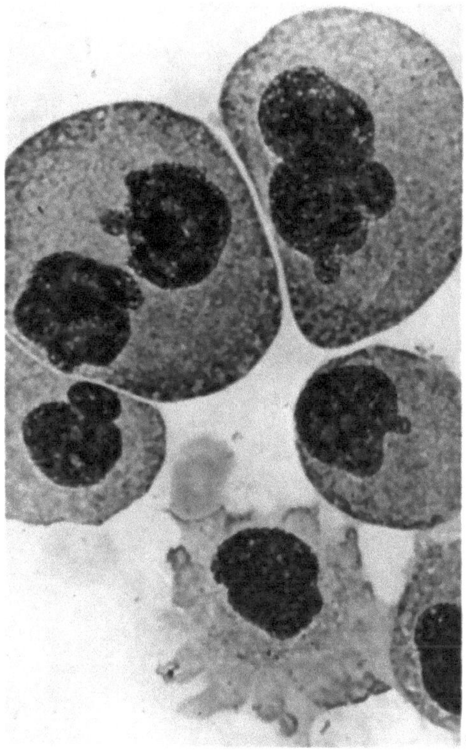
d

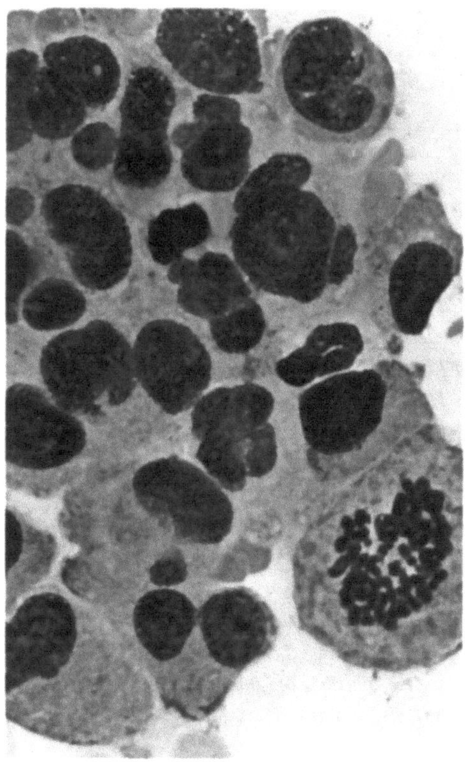
e

Plate 30

Polymorphic pinealoma. Same patient as on page 76/77. Interference phase microscopy.

Clear demonstration of nuclear membrane and nucleoli.

a Lightly structured cytoplasm next to the nucleus. Homogeneous cytoplasmic extrusions in the periphery. × 1,250.

b Tumor giant cell. Large number of nuclei in a rosette formation. × 1,250.

c Mitosis. Chromosomes distributed without apparent order over a large area of the cell. × 1,250.

d Tumor cells in various sorts of loose organization. × 1,250.

e Dense cluster of tumor cells. Above right, mitosis. × 500.

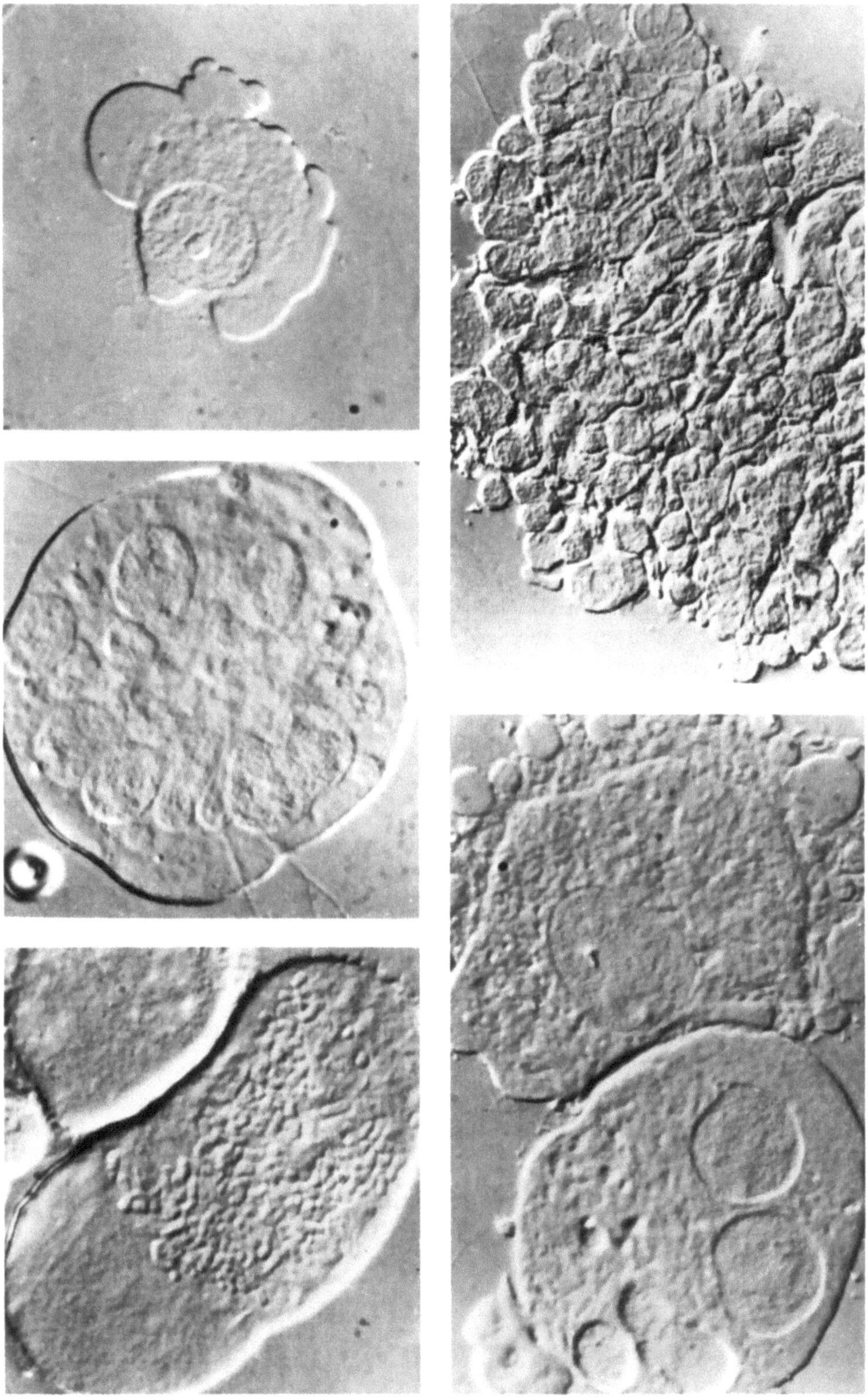

a b c d e

13.1.4. Meningioma, Neurinoma, Primary Sarcomas of the Brain, Hypophyseal Adenoma, Craniopharyngioma, Dermoids

Meningioma, Neurinoma, Primary sarcomas of the brain. Meningeoma and Neurinoma are usually benign tumors, with the result that no tumor cells can be found in the CSF. These tumors do not tend to exfoliate living cells into the CSF, because their rate of mitosis is low and their tissue hard and tightly knit. Cells or cell clusters which appear in the CSF and which do not apparently belong to the physiologic range ought not to be attributed to these tumors.

Some reports have claimed identification of meningioma cells or cell clusters [20, 22, 42, 119], but WIECZOREK [227] reported never finding tumor cells in the CSF in 104 cases of meningioma. The rare malignant forms of these tumors are, of course, exceptions [74, 85]. There are few differences between these cells and those which are almost always present in the equally rare primary leptomeningeal sarcomatosis. The cells appear singly or in loose clusters and are conspicuous for their large size, prominent nucleoli and homogeneous cytoplasm, which may range from strong basophilia to extreme paleness. It is possible to mistake the cells for those of glioblastoma or stem cell leukemia, although the latter generally do not produce giant cells and always demonstrate less cytoplasm.

Hypophyseal Adenoma, Craniopharyngioma, Dermoids. Reports of tumor cells in hypophyseal adenoma and craniopharyngioma are rare [22, 129, 138, 229]. Some authors have described the hypophyseal adenoma as containing isomorphic epithelioid cells with one or two round-to-oval nuclei, which are rich in chromatin.

Tumor cells are also rare in dermoids. Here again, the exceptions are the rare malignant forms, which can exfoliate a wide variety of tumor cells in the CSF, depending on the malignant cell type.

Plate 31

Malignant meningioma. Patient: female, 54 years old. Relapse of a tentorium meningioma. CSF: 10 cells/mm^3, 40 mg% protein.

- a *Large group of polynuclear tumor cells. Minimal accompanying pleocytosis.*
- b *In the center, a hyperchromatic, multinuclear giant cell. Onion-ring structure partially visible.*
- c *Tumor cells with one or two nuclei. Conspicuously large nucleoli.*
- d *4 months later, after operation and radiation therapy. CSF: 1 cell/mm^3, 60 mg% protein. Individual cells, as well as loose groups. Cytoplasmic extensions resembling pseudopodia. Polymorphic nuclei and large nucleoli.*

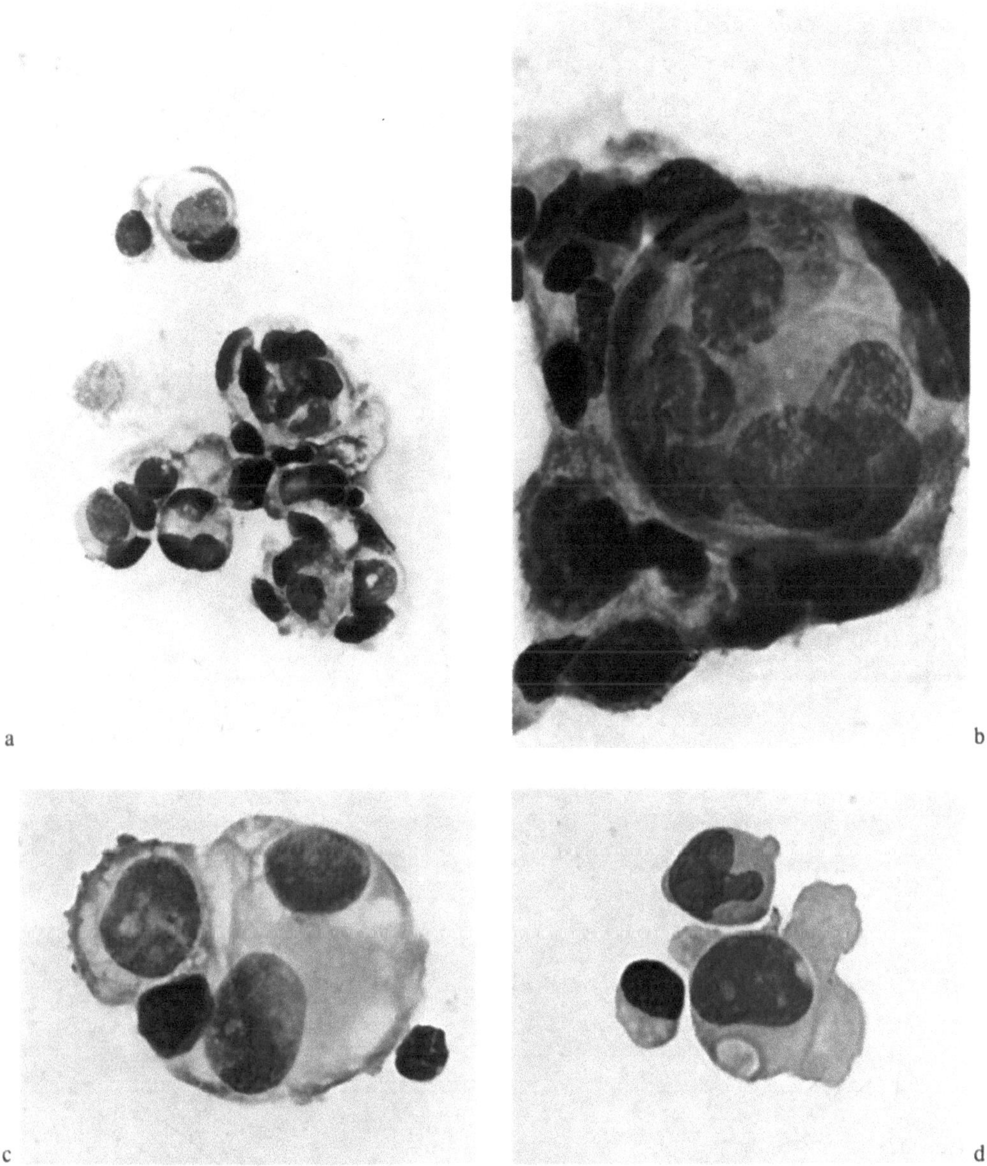
a
b
c
d

Plate 32

a–d Primary leptomeningeal sarcomatosis. Patient: female, 58 years old. CSF 120 cells/mm^3, 120 mg% protein

Groups of tumor cells of various sizes. Some cells strongly resemble transformed monocytes. Cytoplasm predominantly homogenous, generally weakly stained. Intense staining in some cases (a). Cytoplasm of the large cells with fine to middle-sized vacuoles. Nuclear-cytoplasmic ratio altered only in the small cells. Some eosinophils.

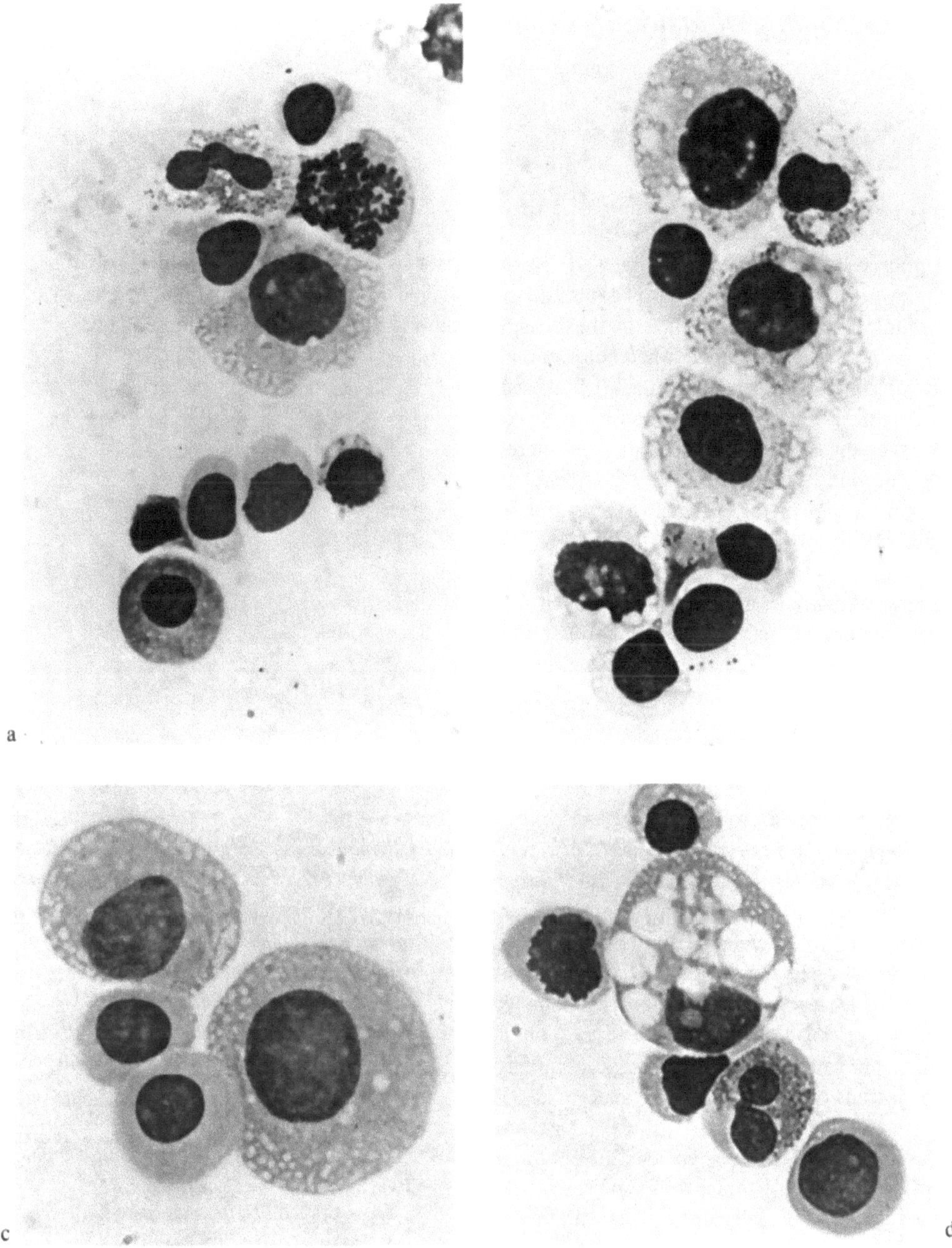
a
b
c
d

13.1.5. Spongioblastoma, Oligodendroglioma, Astrocytoma

Tumor cells appear rarely in the CSF in all three autochthonous tumors. It is doubtful that tumor cells are ever identified in the benign, isomorphic *spongioblastoma* when conventional staining methods (e.g. Pappenheim) are used. One can hardly differentiate between cells of this type and those which appear in CSF altered by infection. Tumor cells may be identified in semi-malignant or malignant spongioblastoma, but this form is rare [227].

Oligodendrogliomas are also comparatively benign tumors, though metastases through the CSF and dissemination of tumor cells in the meninges have been reported [20, 22, 35, 124]. These findings are rather exceptional. The cells of oligodendrogliomas have been described as displaying a large, light pink or blue cytoplasm and often a round, isomorphic nucleus with a moderate amount of chromatin [42]. However, one can certainly identify cells with nuclear polymorphism and hyperchromatism. It is generally difficult to differentiate among tumor cells. ependymal or leptomeningeal cells and monocytes, which are present in increased number and in regressive or hyperplastic form. Tumor cells sometimes form loose clusters, a finding which leads one to suspect malignancy.

The same is true of *astrocytoma* of grades I and II. The isomorphic tumor cells lack cytologic characteristics of malignancy. They cannot be identified as tumor cells when they appear singly; a diagnosis of malignancy is possible when cell clusters are evident [42]. Identification of the tumor cells is easier in the polymorphic astrocytomas, since a shift in the nuclear-cytoplasmic ratio, the large size of the cell, high chromatin content of the nucleus, and cytoplasmic basophilia indicate malignancy. If cytologic characteristics of malignancy are more obvious, a diagnosis of glioblastoma multiforme (astrocytoma of grade IV) [90] is more likely.

Plate 33

a–d *Glioma of the corpus callosum. Patient: female, 61 years old. CSF: 5 cells/mm³, 144 mg% protein. Large cells resembling transformed monocytes, possibly of tumorous origin. Several large nucleoli in each nucleus.*

e *Oligodendroglioma. Patient: male, 40 years old. CSF: 22 cells/mm³, 140 mg% protein. Mass of large, probably tumorous cells with conspicuous nuclear polymorphism and hyperchromatism.*

f *Polymorphic astrocytoma. Patient: female, 50 years old, preoperative. CSF: 7 cells/mm³, 65 mg% protein. Group of suspected tumor cells with dissolving cytoplasm, thick nuclear membranes, and prominent nucleoli.*

g, h *Polymorphic astrocytoma, same patient as (f) postoperative. 3 months later. CSF: 210 cells/mm³, 110 mg% protein. Large, predominantly isolated, polymorphic cells, possibly of tumorous origin. Cytoplasm in process of dissolution.*

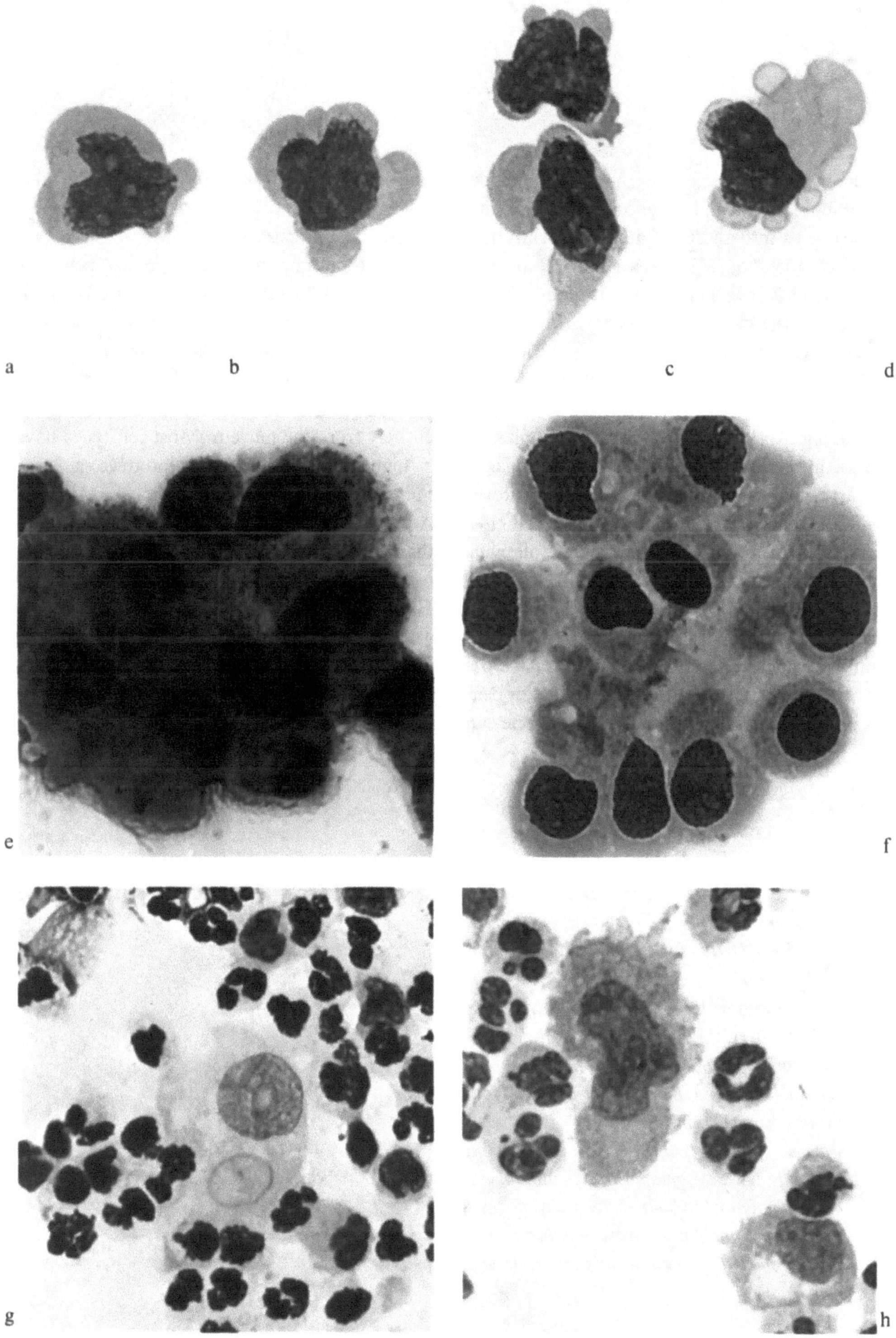
a
b
c
d
e
f
g
h

13.1.6. Glioblastoma Multiforme

Among the autochthonous brain tumors, the medulloblastoma and the glioblastoma exfoliate tumor cells into the CSF most frequently [35, 118, 137, 165, 226, 237]. The polymorphism of the glioblastoma is well known to histologists [238]. Similarly, the cytologic characteristics of glioblastomas are so varied that it is difficult to reduce them to a common denominator.

One type of cell has the following features: a pronounced shift in the nuclear-cytoplasmic ratio, hyperchromatic nucleus and cytoplasm, bizarre nuclear form, a high rate of mitosis, and a tendency to appear in compact cell clusters [175]. Another type shows the following characteristics: extremely large size, varying degrees of shift in the nuclear-cytoplasmic ratio [86, 118, 119], inconsistent affinity for stains, multiple nuclei, very large nucleoli, a loose chromatin texture, and a tendency to appear as isolated individual cells. In addition, one can find cells which hardly differ from those in stem cell leukemia. A variety of cell types may appear in the same preparation.

The cytoplasmic borders of many cells are blurred, and the nuclear borders are often hard to recognize. The cytoplasm of some cells is apparently very vulnerable. At any rate, there is a tendency on the part of the cytoplasm to develop vacuoles or fraying or to detach itself entirely from the nucleus. Most cell types have a high mitotic index and produce a large number of pathologic cell divisions.

A reactive pleocytosis can reach values of more than 1,000 cells/mm^3 [30], composed equally of granulocytes, lymphocytes, and monocytes. Granulocytes occasionally dominate to such an extent that their presence may be falsely attributed to a brain abscess. Hemosiderin accumulated in phagocytes sometimes indicates a past tumor hemorrhage.

Plate 34

Glioblastoma multiforme. Patient: female, 22 years old. ("butterfly" glioma) CSF (lumbar): 80 cells/mm^3, 95 mg% protein.

- *a* *Loose group of large cells without apparent cell borders. Polymorphic, well-defined, sometimes bizarrely shaped nuclei. Some granulocytes and many macrophages containing hemosiderin.*
- *b–d* *Various tumor giant cells, some with extremly wide cytoplasm. Cytoplasmic border indistinct in some cells, quite distinct in others. Coarsely granulated nucleus with distinct nuclear membrane and large, prominent nucleoli.*

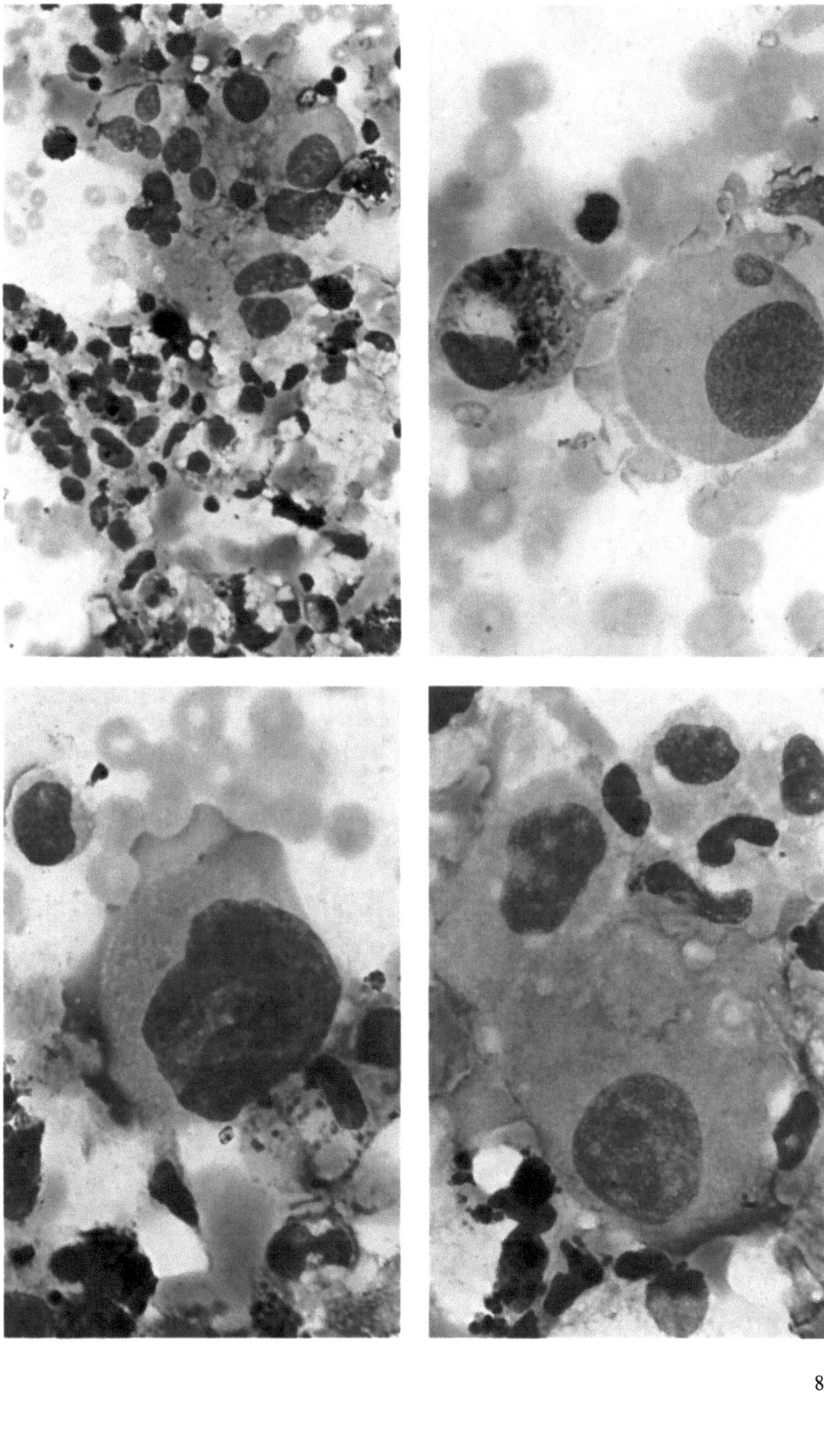
a b c d

Plate 35

a–d *Glioblastoma multiforme. Same patient as on page 81. CSF (ventricular): 440 cells/mm³, 320 mg% protein. (a) Giant cell with many well-defined nuclei of various sizes. Many nucleoli. Pleocytosis caused primarily by neutrophils. (b–d) Papanicolaou stain. The tumor cells show a varied response to Papanicolaou's stain. The cytoplasmic borders are indistinct. The nuclei are sometimes recognizable only as cloudy structures.*

e *Glioblastoma multiforme. Patient: male, 48 years old. CSF: 45 cells/mm³. Single tumor cell.*

f *Glioblastoma multiforme. Patient: male, 54 years old. CSF: 12 cells/mm³, 54 mg% protein. Cluster of tightly grouped, well-defined hyperchromatic nuclei. Only remnants of cytoplasm.*

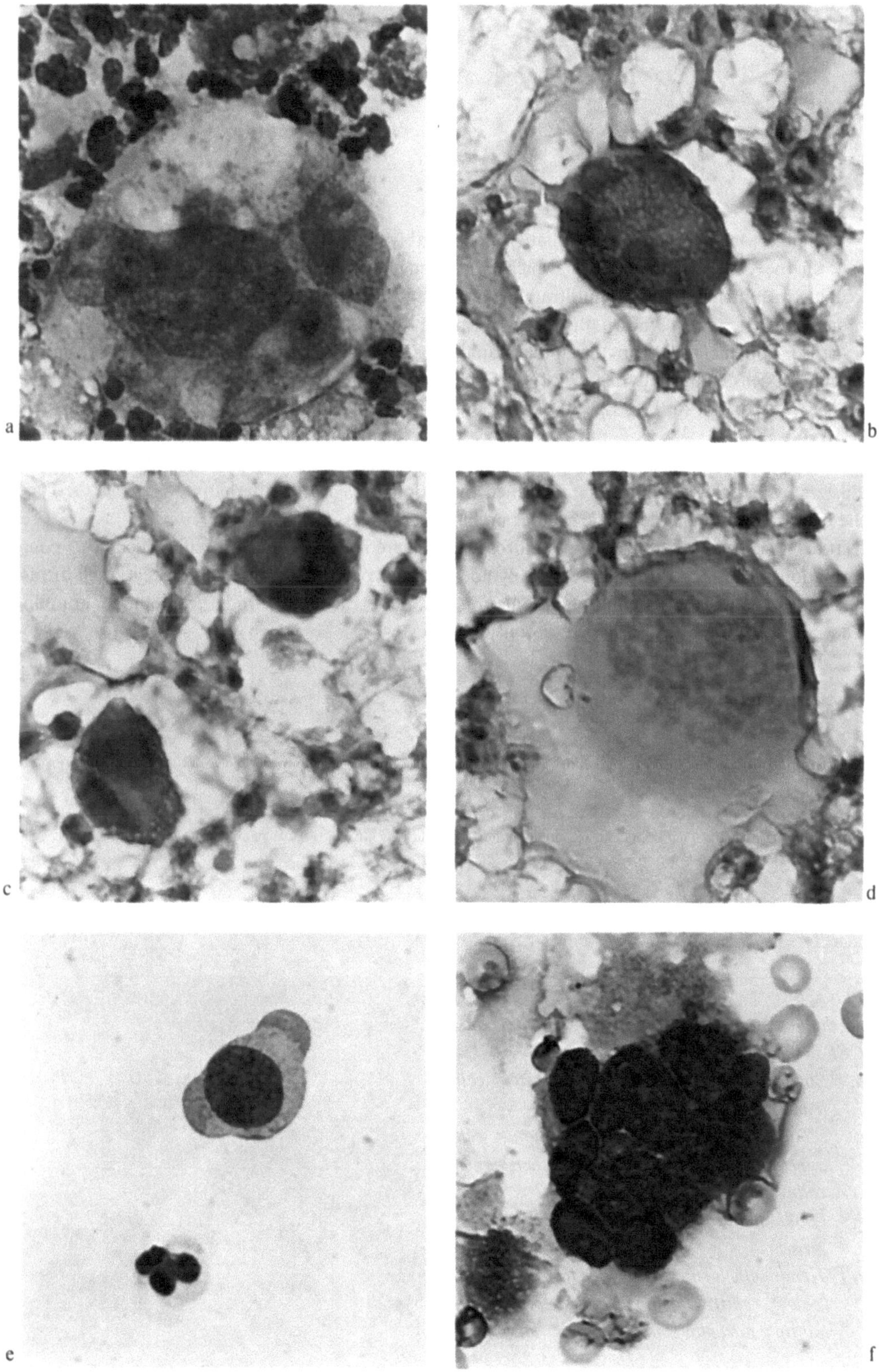
a
b
c
d
e
f

13.1.7. Medulloblastoma

Medulloblastomas frequently exfoliate tumor cells in the CSF, a characteristic which is in keeping with their tendency to infiltrate the leptomeninges. It is well known that they form so-called drip metastases in the spinal canal. Malignant cells can usually be identified in the preparation [20, 35, 46, 115, 122, 124, 196, 216].

One finds both individual cells and many cell clusters, which can sometimes be identified by the recurrence of a characteristic cellular architecture—the true rosette and pseudorosette formations. The nuclei are markedly polymorphic and sometimes stain with fine granulation. They are sometimes turnip-shaped or considerably elongated out and have one or more large nucleoli. The cytoplasm is usually homogeneous, and the cell is round when isolated but appears polygonal when found in clusters.

Cytoplasmic basophilia varies greatly from cell to cell, and inclusions or vacuoles are rare. Isolated cells sometimes resemble those in stem cell leukemia [11, 193]. Primary leptomeningeal sarcomatosis deserves consideration as an additional diagnostic alternative.

There is usually an unspecific pleocytosis composed of lymphocytes and monocytic cell forms. One occasionally finds eosinophilic granulocytes.

Plate 36

Medulloblastoma. Patient: male, 6 years old. CSF: 40 cells/mm³.

- *a, b* *Single tumor cells. Similarity to stem cell leukemia.*
- *c* *"Tumor cell phagocytosis". A tumor cell appears in the cytoplasm of another tumor cell.*
- *d* *Loose tumor cell group. Bizarre nuclear forms.*
- *e, f* *Tumor cell cluster. Some cells form a rosette (below right). In most cases, prominent multiple nucleoli, some extremely large.*

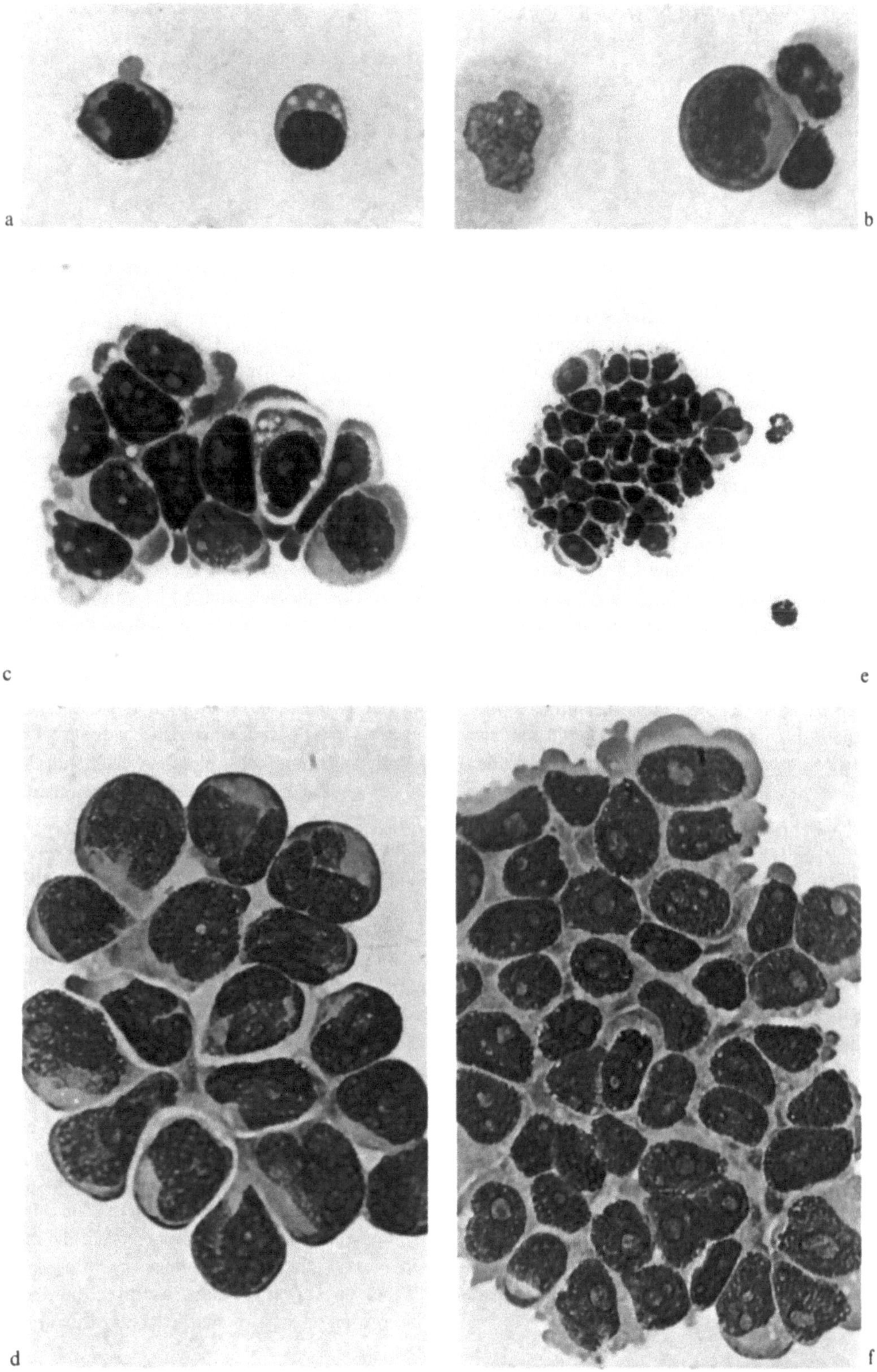
a
b
c
e
d
f

13.2. Metastases in the CNS

Certain tumors have a pronounced tendency to metastasize to the brain, but the mere existence of metastasis is, in itself, indicate of a high degree of malignancy on the part of the tumor. Therefore, it is not surprising that tumor cells are more frequently found in cerebral metastases than in primary tumors of the CNS. Tumor cells can be identified in the CSF in 30–40% of cerebral metastases. Some authors report a more limited [35], others a much higher rate of success [20, 129]. No matter how small the number of cells, a slide preparation must be made and carefully examined if a tumor is suspected. One can sometimes diagnose carcinoma metastases with only a few cells or a single cell cluster which fulfills the criteria of malignancy.

It is known that some tumors metastasize extremely early to the brain (melanoma), while others show an unexpectedly limited tendency to do so (stomach carcinoma). However, tendency to develop cerebral metastasis does not necessarily correlate with frequency of positive tumor cell findings in CSF.

Tumor cells of epithelial provenance have a conspicuous cytoplasmic structure and are larger than cells found in glioma, medulloblastoma, sarcoma and the hemoblastoses. The cytoplasm often stains inhomogeneously, and the presence of irregular vacuoles suggests cytoplasmic inclusions. Structures may vary considerably from cell to cell within the same slide. The cytoplasmic inclusions develop either through a process of denaturation or as a result of independent cellular function. Phagocytic inclusions are evidence against a diagnosis of malignancy. Cytoplasmic differentiation reflects the degree of differentiation of the carcinoma itself. Undifferentiated cytoplasm in a cell picture produced with panoptic stain allows few conclusions as to the nature of the tumor.

One distinguishes among three types of carcinoma according to histologic criteria: the epidermoid, the glandular, and the undifferentiated form. One can sometimes recognize some of these characteristics in CSF cell preparations, so that diagnoses are possible.

The *epidermoid type* can be subdivided into two categories, according to the presence or absence of cornification. Epidermoid carcinoma with cornification has a limited tendency to metastasize, so that tumor cells appear very rarely in the CSF. If they are found, identification is relatively easy, since their cytologic characteristics are conspicuous and quite constant. The cells usually appear singly, though occasionally in clusters, and both giant cells and small cells can be found. The nuclear-cytoplasmic ratio may be slightly increased, if at all. The presence of specific cell types is characteristic: tadpole cells, spindly squamoid cells, birds-eye cells and so-called "ghost cells". The latter derive their name from the fact that they contain the merest suggestion of a nucleus or no nucleus at all. The most significant cytologic characteristic is keratin production, which may be demonstrated in yellow or brilliant orange tones with Papanicolaou stain or as deep violet with Pappenheim stain. Keratin often gives the cytoplasm sharp contours. Mitosis is rare.

The less differentiated squamous cell carcinoma has a greater tendency to metastasize, and cells of these cancers are occasionally found in CSF. In most cases, these derive from metastases of bronchial carcinoma. The cells may appear singly but are most often found in clusters, which are usually organized in a mosaic pattern. The cells are sometimes packed together or piled one on top of the other. Cell outlines are often indistinct or appear to have a frayed border. The cytoplasm has no characteristic structures and is sometimes so basophilic that the nuclear borders are almost indistinguishable. The nuclear-cytoplasmic ratio is almost always increased, and the nucleus has a distinct membrane and usually contains several prominent nucleoli. Mitosis is not uncommon.

Cells of the *glandular type* appear more frequently in clusters in which cells overlap and give the appearance of a gland. The latter formation is rarely found in CSF. Many sorts of tumor are conspicuous for the fact that they do not appear in solid clusters but rather in loose groups or as single cells. Individual cells can attain extreme size in the CSF (over 200 μ). The gigantic cytoplasmic body has amoebic extensions and is strongly basophilic with Pappenheim stain. Multiple nuclei appear often. They usually take an excentric location and have—as a special characteristic—a single (rarely multiple), prominent, extraordinarily large nucleolus (62.). Many of these tumor cells produce mucus, which is deposited in the cytoplasm in large vacuoles or in a wreath of vacuoles in the periphery of the cell. It often happens that a single bubble of mucus occupies the greater part of the cell body, pushing the nucleus to the border; this is the signet-ring cell. Other tumors occasionally produce cells which appear in the CSF and which may look like signet-ring cells. Differentiation from signet-ring cells of the proliferating mononuclear phagocytic system must be made. However, the presence of signet-ring cells and other criteria of malignancy suggest the existence of an epithelial tumor capable of producing mucus. Many adenocarcinomas metastasize very early, so that one most often expects positive cytologic findings in those growths.

The *undifferentiated type* of carcinoma may demonstrate either isolated cells or loosely organized cell clusters. They occasionally appear to be more compact, because the cells have extremely pliable cytoplasm and nuclei, which allows the cells to fit together without gaps. This sort of pliability, which can be observed in cells in the CSF, entails great variety in the shapes of nuclei and cytoplasm, variety of a sort that is quite unusual in such tumor cells. All criteria of malignancy are fulfilled. The nuclear-cytoplasmic ratio is significantly increased, so much so that one can sometimes find only a narrow cytoplasmic border. The large, strongly hyperchromatic nucleus contains one or several large nucleoli and tends to dominate the cytoplasm. The latter is quite vulnerable and takes a strongly basophilic stain.

Neutrophils, RBCs and evidence of phagocytized RBCs often accompany such tumor cells. Cytologic classification of undifferentiated carcinoma is difficult, since similarities to autochthonous brain tumors—especially to glioblastoma and medulloblastoma—and to hemoblastoses are possible.

13.2.1. Carcinomas of the Respiratory Tract

Over 30% of all metastases in the brain originate in carcinomas of the respiratory tract (usually bronchial carcinoma), and, in fact, tumor cells in the CSF most frequently indicate metastases of such tumors [66, 112, 137, 149, 199, 213]. Three forms of bronchial carcinoma can be differentiated according to histologic criteria: adenocarcinoma, undifferentiated carcinoma (oat cell carcinoma), and epidermoid carcinoma. The latter can be divided into a squamous type with cornification and a large cell type without cornification. The epidermoid carcinomas are the most common types followed by the oat cell carcinomas.

Cells of the oat cell carcinoma and the glandular carcinoma appear most frequently in the CSF, even if the primary tumor has not yet been discovered. Epidermoid carcinomas have a more limited tendency to metastasize, so that cells of these tumors are less often identified in the CSF. If the tumor has retained the capacity for cornification, then the probability of finding tumor cells in the CSF is even more limited as a result of the relative benignity of the neoplasm.

Plate 37
Bronchial carcinoma. Patient: male, 54 years old. Oat cell carcinoma. CSF: 12 cells/mm^3, 60 mg% protein. Cytologic characteristics do not indicate the provenance of the tumor.
a Cells tightly grouped in the form of a palisade. Pronounced nuclear polymorphism.
b Telophase of quadrupolar mitosis.
c Cells with large, turnip-shaped, elongated nuclei, rich in chromatin. Above: Telophase of mitosis.
d Compact cluster of tumor cells. Below: atypical cell division.

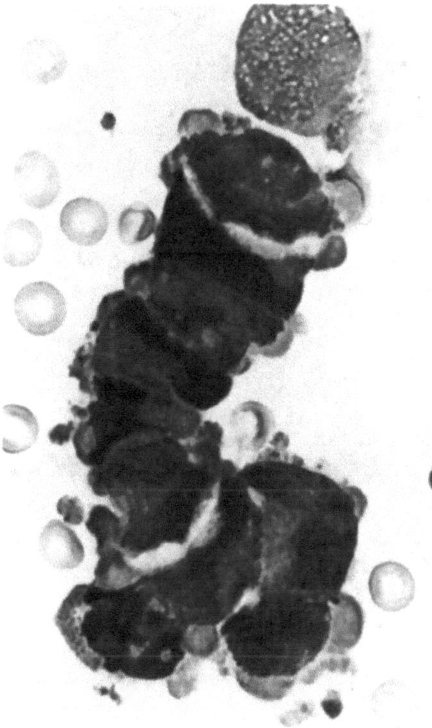

a

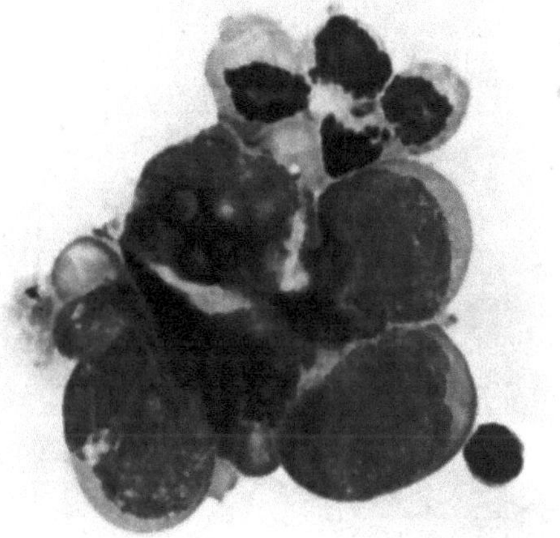

b

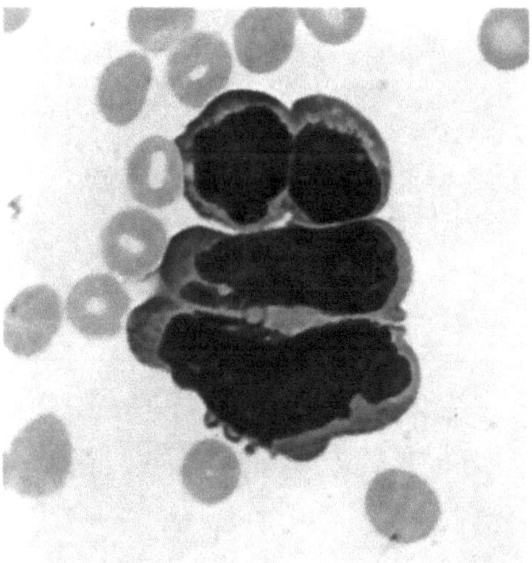

c

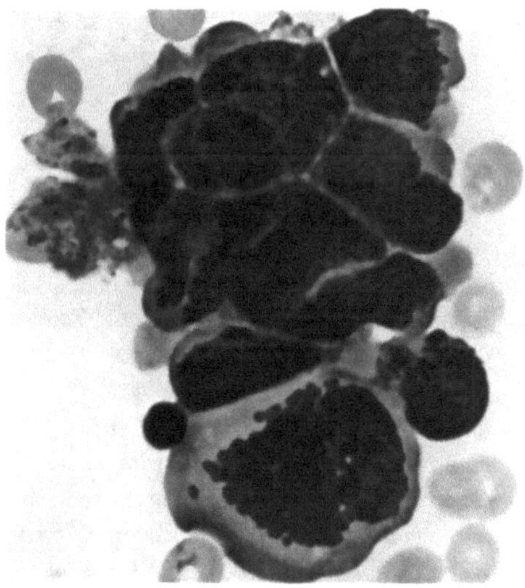

d

Plate 38

a, b *Bronchial carcinoma. Patient: male, 42 years old. Adenocarcinoma. CSF: 10 cells/mm^3, 65 mg% protein. (a) Two tumor cells. Nuclear polymorphism. Cytoplasmic border interspersed with vacuoles ("string of pearls" organization). (b) Tumor cell cluster with a tendency towards glandular organization. Relatively isomorphic nuclei.*

c, d *Bronchial carcinoma. Patient: male, 66 years old. Pulmonary adenocarcinoma. CSF: 60 cells/mm^3, 90 mg% protein. (c) and (d) Tight cluster of very large cells. (c) Cytoplasm interspersed with large vacuoles. (d) Nuclear crowding.*

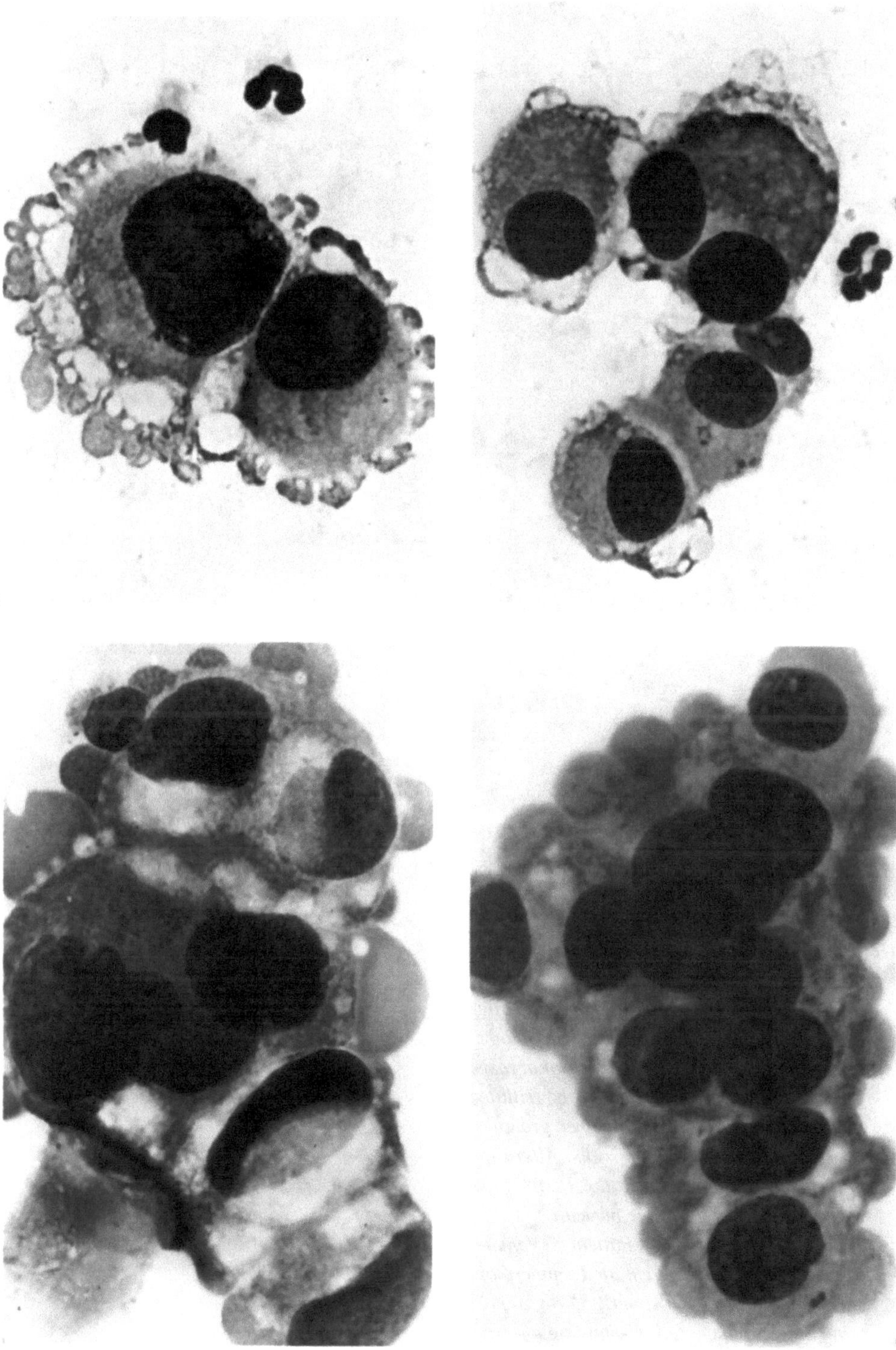

a b c d

Plate 39

a–c *Bronchial carcinoma. Patient: male, 50 years old. Undifferentiated epidermoid carcinoma. CSF: 3 cells/mm^3. (a) Autoradiography after in vitro incubation with tritiated thymidine. Loose epithelioid cell group. Anisokaryosis. Many cells clearly labeled with tritiated thymidine. Focus in plane of silver granules. (b, c) Tight cluster of tumor cells. Alteration in the nuclear-cytoplasmic ratio. Distinct nuclear membranes. Multiple nucleoli.*

d *Bronchial carcinoma. Patient: female, 62 years old. Undifferentiated epidermoid carcinoma. CSF: 4 cells/mm^3, 35 mg% protein. Single tumor cell (below right) with strong cytoplasmic basophilia. No other indication of the provenance of the cell.*

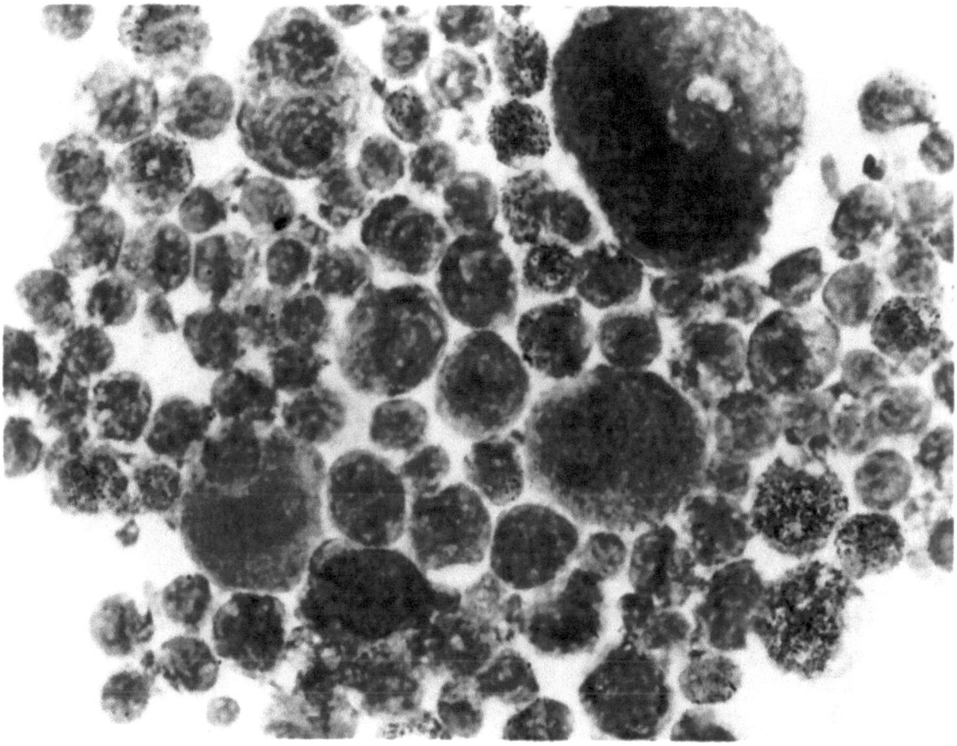

a

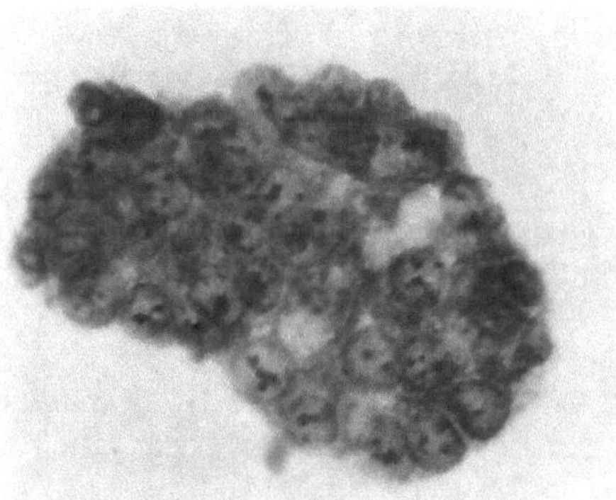

b

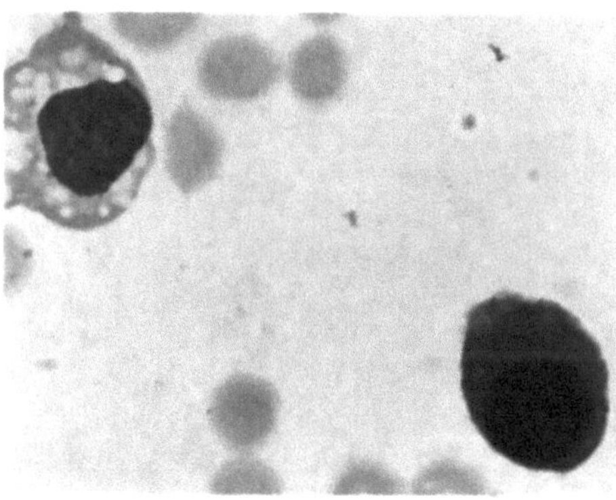

d

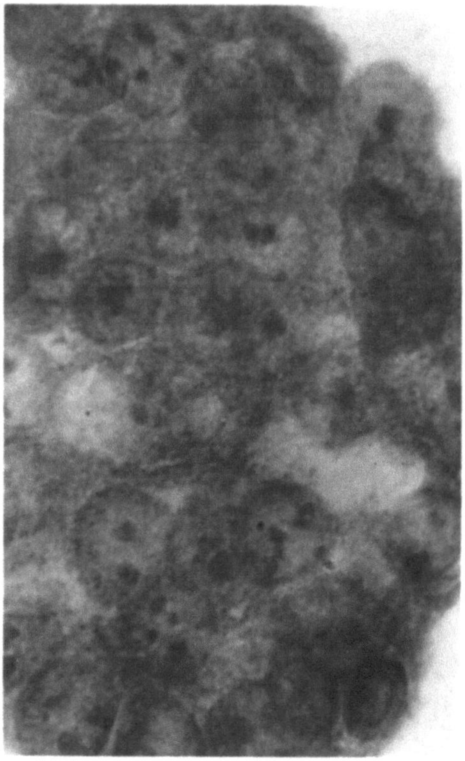

c

Plate 40

Keratinizing epidermoid carcinoma. Patient: male, 52 years old. Penetration of keratinizing epidermoid carcinoma from the cells of the cribiform plate into the meninges. CSF: 320 cells/mm^3, 90 mg% protein.

a Tumor cell group. No alteration in the nuclear-cytoplasmic ratio.

b Tumor cell group. Keratinization demonstrated by the orange stain of the cytoplasm. Papanicolaou's stain.

c Tadpole cell.

d Single tumor cell in mitosis.

e Polynuclear tumor cell with strongly basophilic cytoplasm.

f Epithelial cell without definite signs of malignancy. Leukocyte invasion of the cytoplasm.

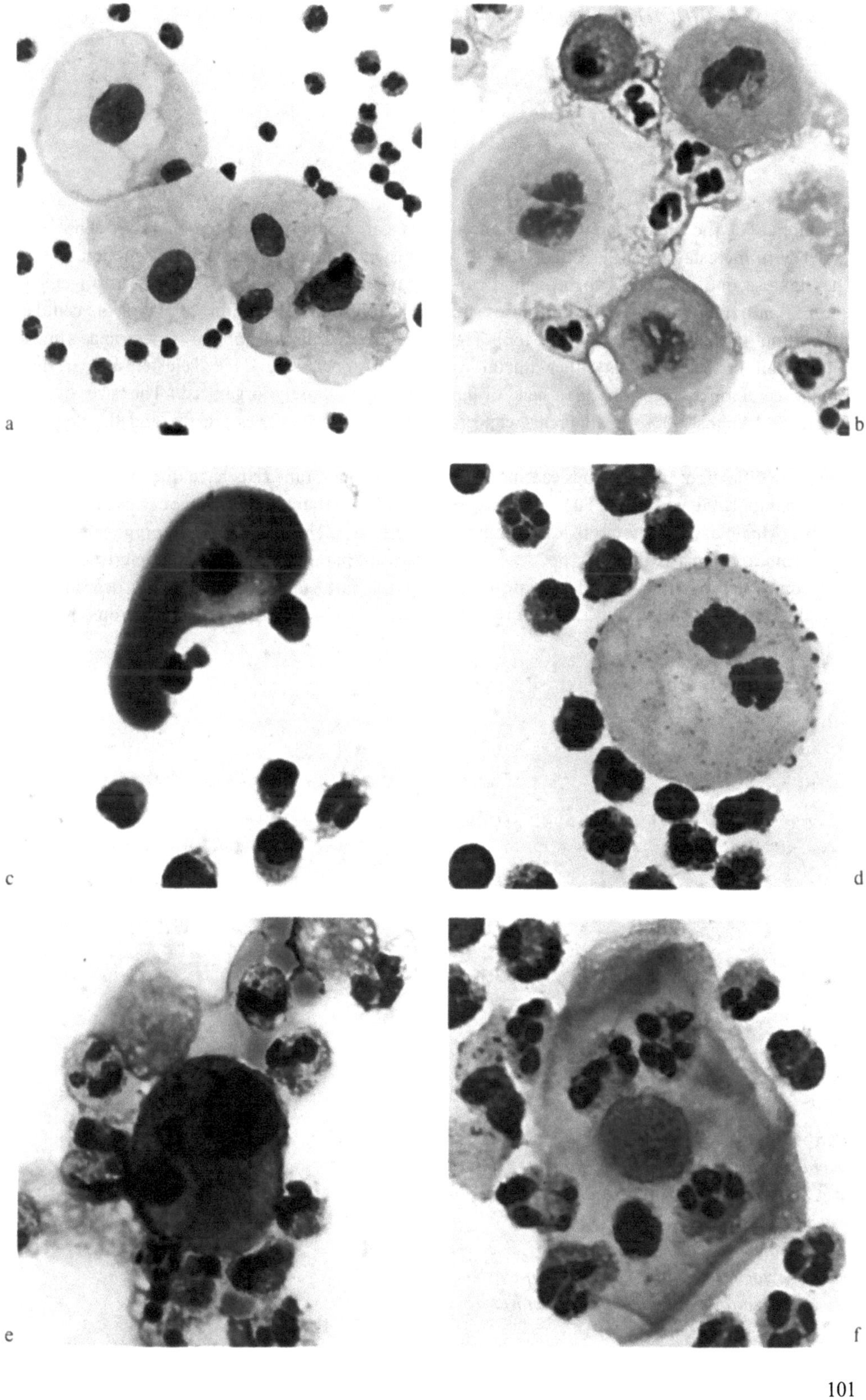
a
b
c
d
e
f

13.2.2. Carcinomas of the Breast

Carcinomas of the breast show a pronounced tendency to metastasize to the CSF [112, 149]. They take second place only to bronchial carcinoma, and tumor cells from metastases are often found in the CSF [37, 42, 137, 166]. The tumors can be classified as either ductal or lobular carcinoma according to their origin [8, 31, 33]. Almost 90% of all breast cancers are ductal carcinomas. On the basis of a histological classification, the scirrhous carcinoma is the most important, with about 80% of all breast cancers. Medullary carcinoma is less common, and adenocarcinoma is rare.

It is sometimes possible to differentiate cytologically among the various types of tumor. Ductal carcinoma most often displays dense cell clusters, though single cells of moderate size may occasionally appear. Tandem arrangement and blastula formation are typical cellular patterns. Cells in lobular carcinoma almost always appear singly. If cell clusters appear, they are only loosely organized. The size of the individual cells varies greatly, and the nuclear-cytoplasmic ratio is higher than in cells of ductal carcinoma. Some cells have fine to middle-sized droplets in the cytoplasm. One can usually distinguish between these and lipid storage cells on the basis of malignant nuclear characteristics. Cells of the rare adenocarcinoma can demonstrate cytologic characteristics of muciparous tumors.

Plate 41

Ductal carcinoma of the breast. Patient: female, 32 years old. CSF: 23 cells/mm³, 82 mg% protein.

Large number of tumor cells of varying size and nuclear-cytoplasmic ratio. Some cells in clusters. Multipolar mitosis in middle of the picture.

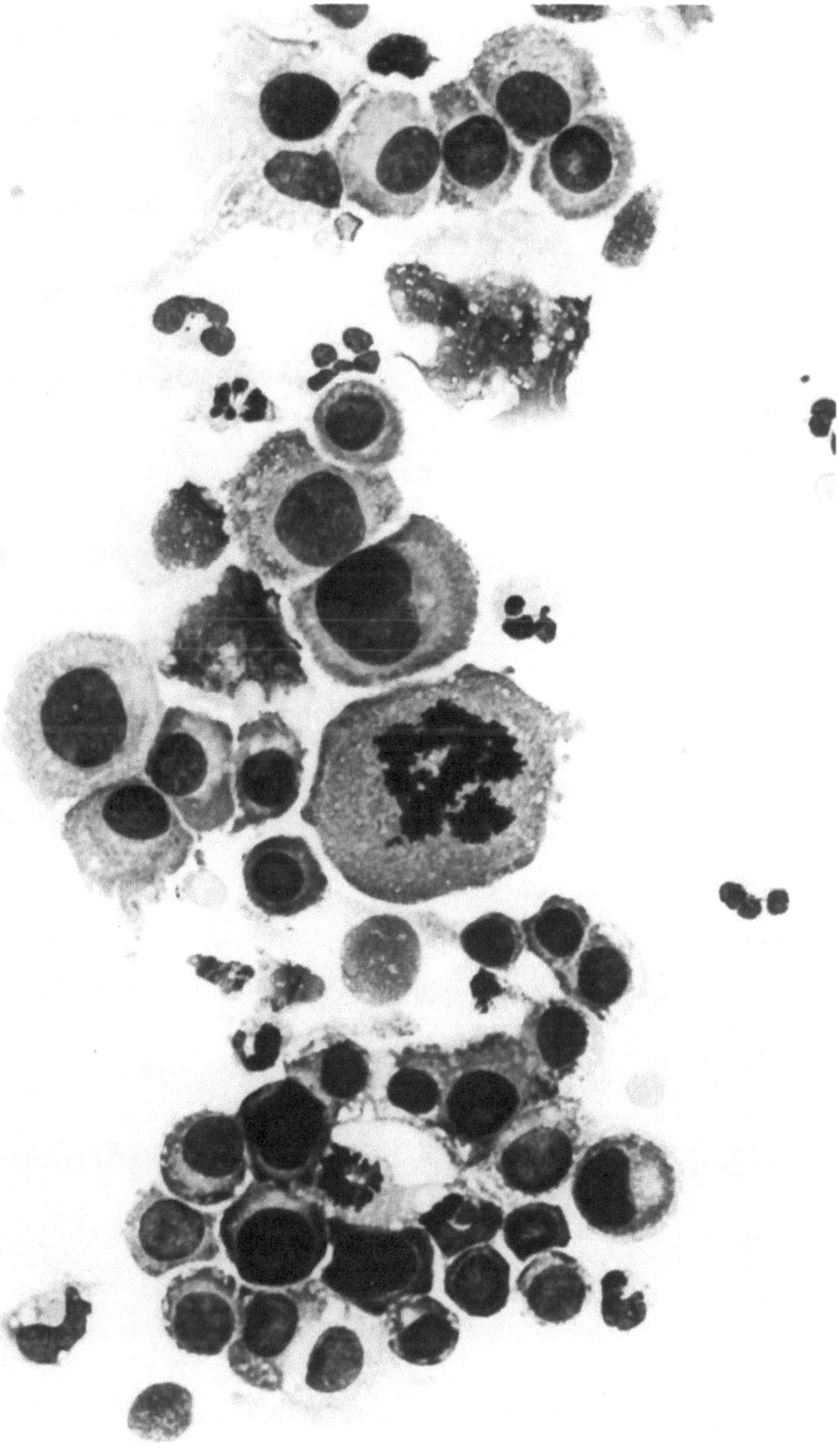

Plate 42

a–c *Lobular carcinoma of the breast. Patient: female, 71 years old. CSF: 7 cells/mm^3. An overwhelming majority of single tumor cells which vary greatly in size and shape.*

d *Adenocarcinoma of the breast. Patient: female, 52 years old. CSF: 6 cells/mm^3. Interference phase microscopy. Conspicuous nucleolus. Morphologic characteristics suggest a muciparous tumor.*

e–g *Ductal carcinoma of the breast. Same patient as plate 41.*
(e) Cell cluster and tandem arrangement. (f) Loosely organized group of tumor cells. Nuclei with 1 or 2 nucleoli and varying affinity for stain. (g) PAS reaction. PAS-positive substances clearly localized in the periphery of some cells, less evident in others.

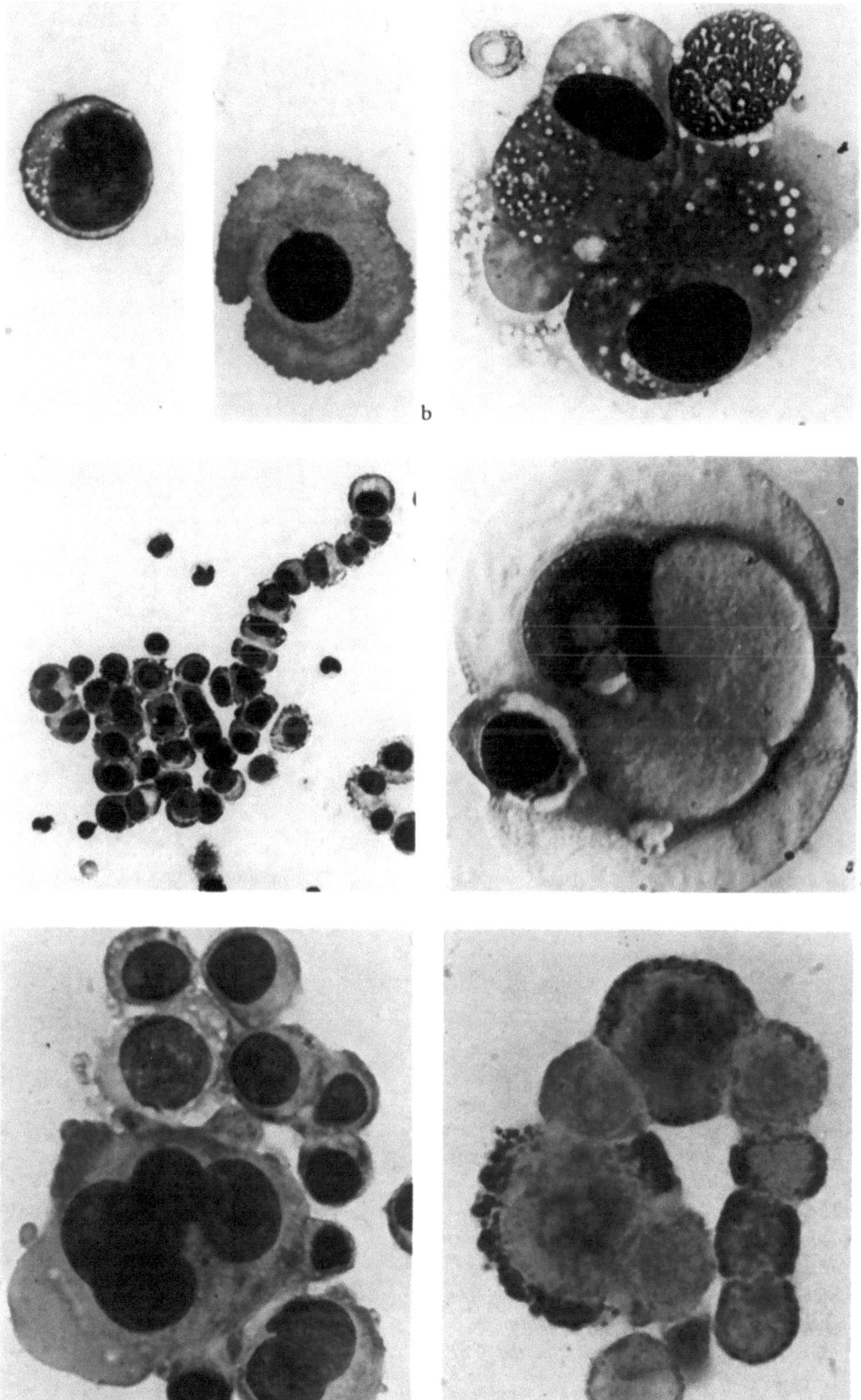

a b c

e d

f g

Plate 43

Carcinoma of the breast. Patient: female, 66 years old. CSF: 13 cells/mm^3.

a Loose cluster of polymorphic tumor cells.

b Tumor cell cluster. One or two nucleoli in each of the coarsely-granulated nuclei.

c Signet-ring cell surrounded by other tumor cells, some with fine droplets in the cytoplasm.

d Compact "floating metastasis" in the CSF. Large mass of tumor cells.

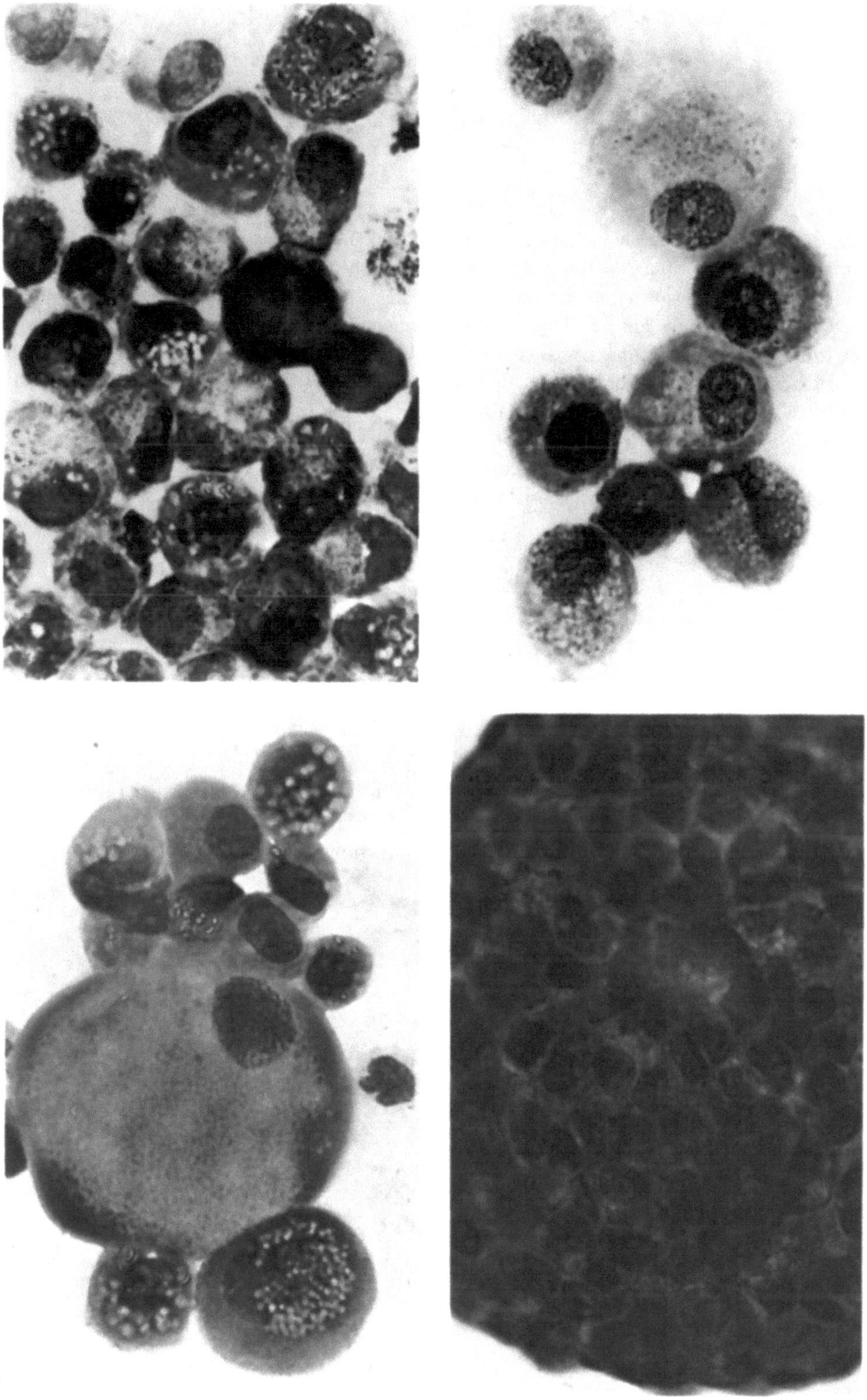

a b c d

13.2.3. Carcinoma of the Alimentary Tract

The stomach carcinoma is the second most common neoplasm after the bronchial carcinoma, though its tendency to metastasize to the CNS is apparently limited. This may be due to the short survival time of patients with stomach carcinoma. The probability of finding tumor cells in the CSF is high if metastasis in the CNS takes place. In fact, it is not rare that a stomach carcinoma is first discovered through CSF cytology after metastasis in the CNS. Two forms of carcinoma can be distinguished according to their histological structure – scirrhous carcinoma and adenocarcinoma. The adenocarcinoma most often produces positive cytologic findings. One sometimes finds wide dissemination of tumor cells as an indication of carcinomatous meningitis, even when a cerebral metastasis cannot be found at autopsy [94, 125].

Plate 44

Carcinoma of the stomach. Patient: male, 33 years old. Adenocarcinoma of the cardia. CSF: 110 cells/mm^3.

a Loose tumor cell groups. Extreme differences in size between tumor cells and leukocytes. × 250.

b Tumor cell group. Good cytoplasmic organization. Cytoplasmic borders interspersed with vacuoles, organized in the form of a string of pearls. In general, one large, prominent nucleolus in each coarsely granulated nucleus. × 1000.

c Autoradiography after in vitro incubation with tritiated thymidine. High concentration of silver grains as a sign of thymidine incorporation during incubation. × 400.

d Fluorescent staining with acridine orange. Tumor cell cluster. Strong red fluorescence of cytoplasm. × 1000.

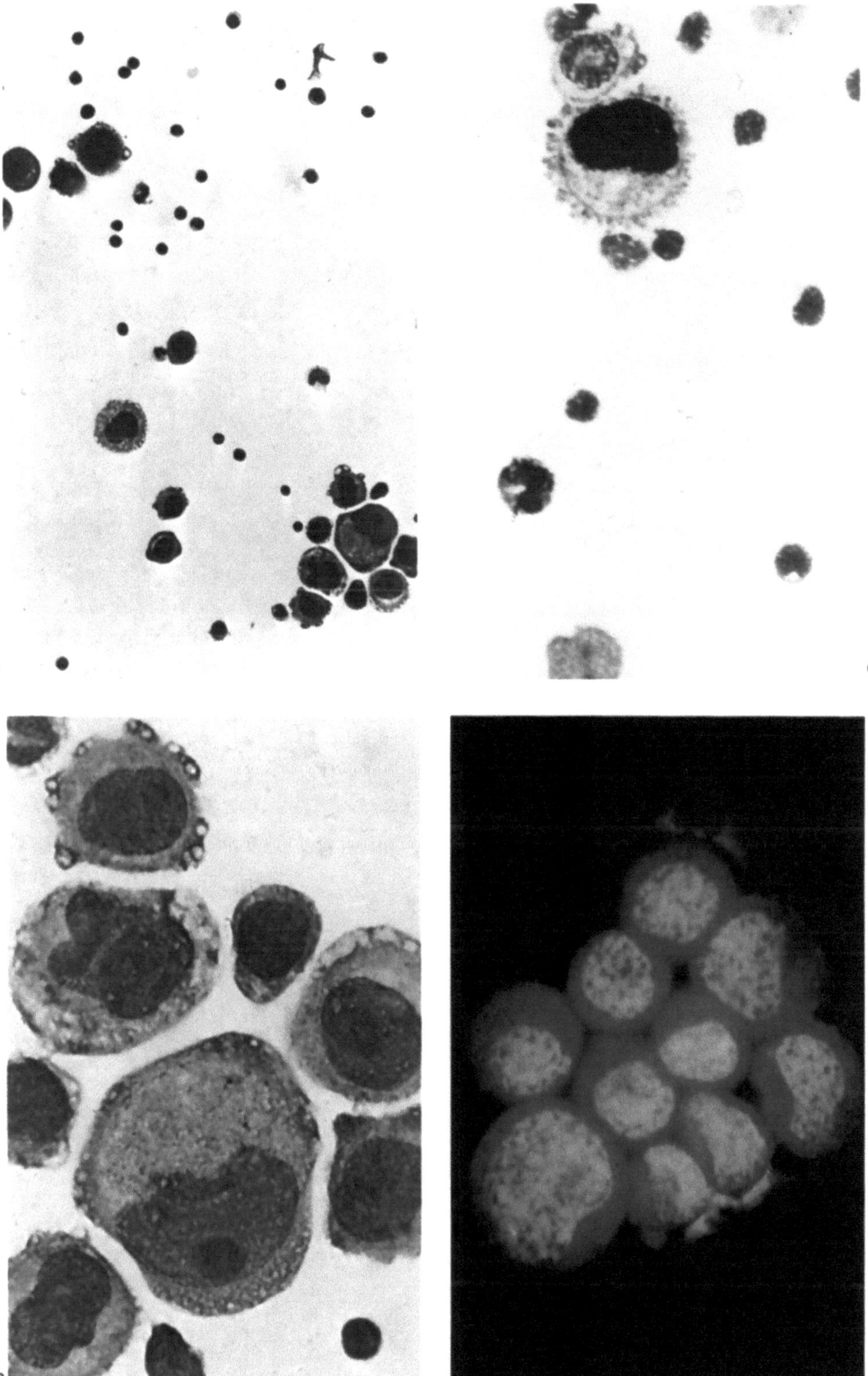

a c

b d

Plate 45

a Adenocarcinoma of the stomach. Patient: female, 70 years old. Adenocarcinoma and herpes zoster meningitis. CSF: 320 cells/mm^3, 85 mg% protein. Autoradiography after incubation with tritiated thymidine. Pleocytosis consisting primarily of granulocytes and lymphocytes. Several cells with strong evidence of thymidine incorporation.

b Adenocarcinoma of the stomach. Same patient as plate 40. Signet ring cell, small tumor cell and lymphocytes.

c Adenocarcinoma of the gall bladder. Patient: female, 54 years old. CSF: 15 cells/mm^3. Signet ring cell, typical of muciparous adenocarcinoma. Nuclei pressed against the cytoplasmic border.

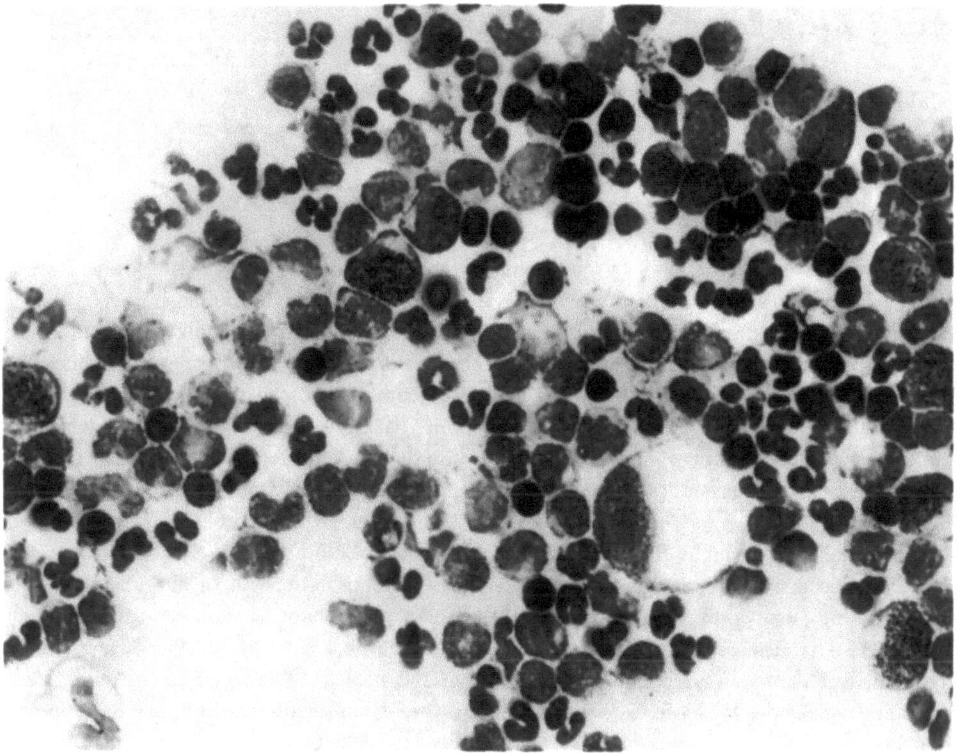

a

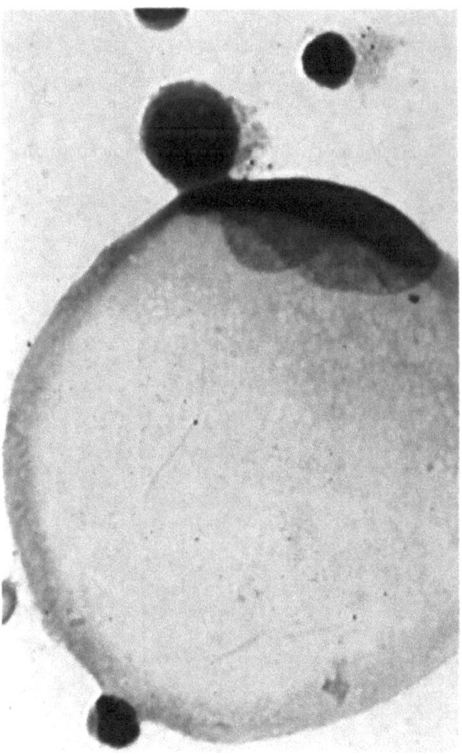

b

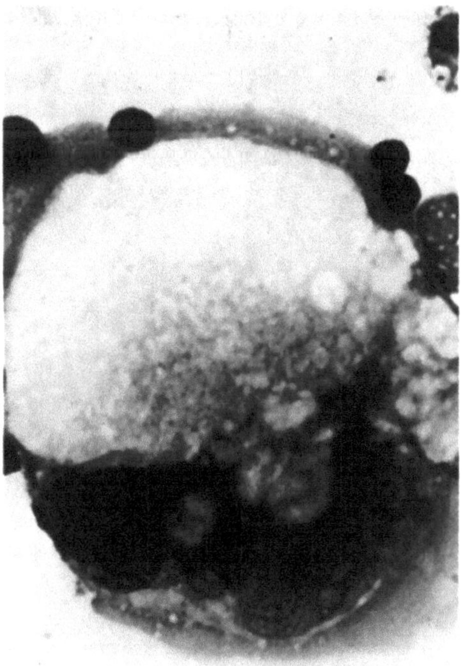

c

13.2.4 Melanoma

Metastatic invasion of the CNS by melanoma is not uncommon. Diagnosis of the extremely rare primary meningeal melanosarcoma is permissable only after a careful search for an extracerebral melanoma. Tumor cells always appear in the CSF in primary meningeal melanosarcoma, while a search for tumor cells is often fruitless in cases of metastatic invasion [6, 41, 115, 166, 195, 208].

Melanoma cells appearing in the CSF are large. They often occur singly or in pairs, though occasionally in small groups or even tightly grouped in a tandem arrangement. Nuclei are round or oval, contain a large amount of chromatin and usually display more than one prominent nucleolus. Mitoses—some of them multipolar—are a common finding. The staining properties of the cytoplasm with Pappenheim stain can vary; it sometimes appears strongly basophilic. The cytoplasmic border is usually fringed with vacuoles. Only a portion of the cells contains melanin pigment and one must sometimes search to find any. It is located in fine, dark-brown or black granules which are usually concentrated in one region of the cytoplasm, though they may occasionally be distributed uniformly throughout.

Ingested melanin appears either in fine granules or in clumps in the cytoplasm of the phagocyte. These cells connot be mistaken for tumor cells since they lack signs of malignancy, particularly in the nucleus. Phagocytized melanin, which cannot be differentiated from hemosiderin with Pappenheim stain, is usually the only cytologic evidence of melanoma metastasis.

Plate 46

Melanoma menigitis. Patient: female, 52 years old. CSF: 120 cells/mm^3, 95 mg% protein, 15 mg% sugar.

- *a Loose group of tumor cells with epithelial structure. On the right a few cells with finely granulated pigment deposits in the cytoplasm.*
- *b Dense cell complex. Nuclei round to oval. Tumor cells predominantly free of melanin, with the exception of one cell (middle right).*
- *c Group of tumor cells. Left: dense pigment deposits throughout the entire cell. Middle: karyolysis.*

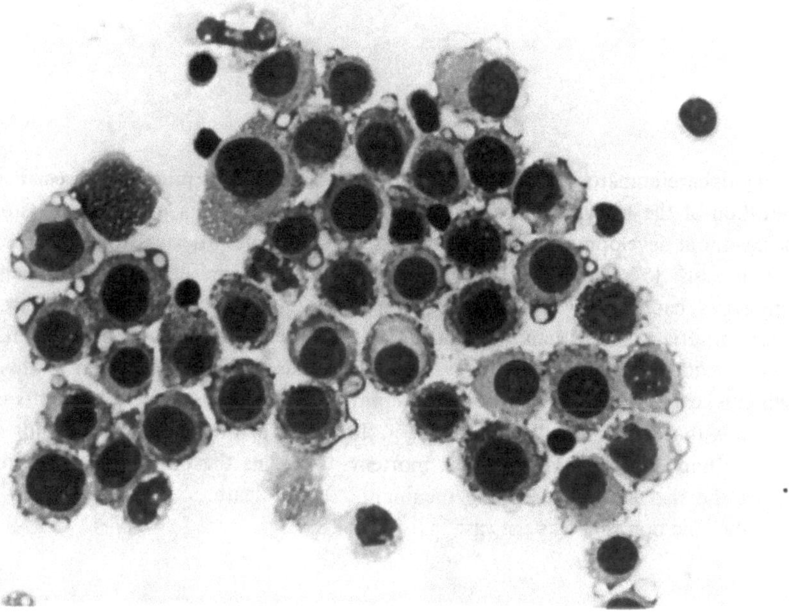

a

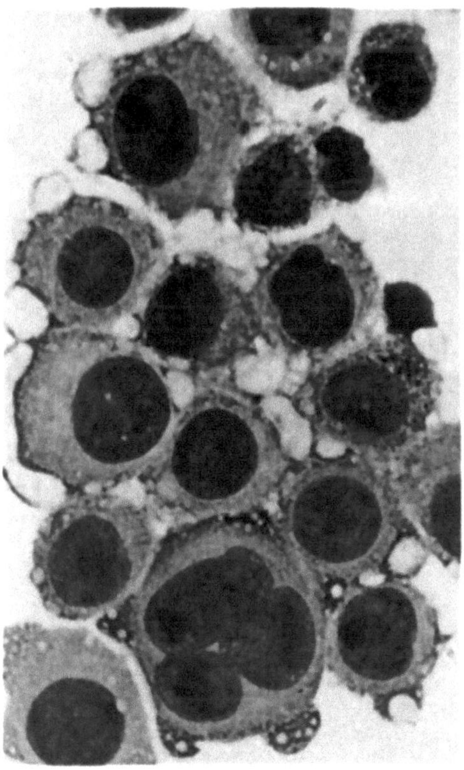

b

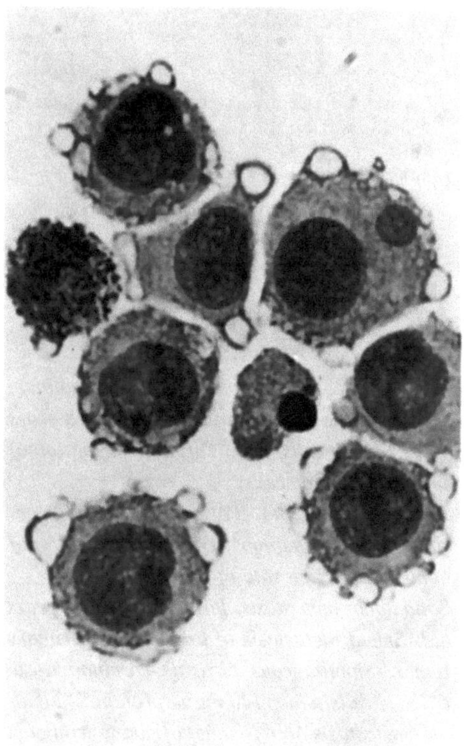

c

13.2.5. Meningitis Carcinomatosa

Meningitis carcinomatosa is a diffuse tumor cell infiltration of the leptomeninges and the CSF space, without development of large metastases in brain tissue [57, 144, 150, 197]. A certain diagnosis is based solely on pathologic anatomical criteria, rather than on clinical and cytologic evidence. Autopsy often reveals both meningitis carcinomatosa and solid cerebral metastases that were not clinically evident. It is usually impossible to ascertain post mortem whether the metastasis caused the meningitis or whether the opposite was the case.

Tumor cells must be present if meningitis carcinomatosa is clinically diagnosed. The cells most often derive from adenocarcinoma, especially from those of the alimentary tract [94, 125], the bronchi, and the breast, in descending order of frequency. Sarcomas may occasionally cause diffuse infiltration of the leptomeninges and the CSF space. Meningitis is often discovered before the primary tumor and sometimes remains the only sign of a neoplastic process until death.

Plate 47

a Meningitis carcinomatosa. Carcinoma solidum simplex of the cervix. Patient: female, 48 years old. CSF: 30 cells/mm^3. Diffuse distribution of epithelioid tumor cells.

b Rhabdomyosarcoma. Patient: male, 44 years old. CSF: 45 cells/mm^3. Diffuse distribution of tumor cells. High rate of mitosis.

c Solid tumor metastasis. Patient: male, 64 years old. Spinal metastasis of an unknown primary tumor. Simultaneous pleuritis carcinomatosa. CSF: 4 cells/mm^3, 140 mg% protein. "Floating metastasis" consisting of many compact tumor cells, some in mitosis.

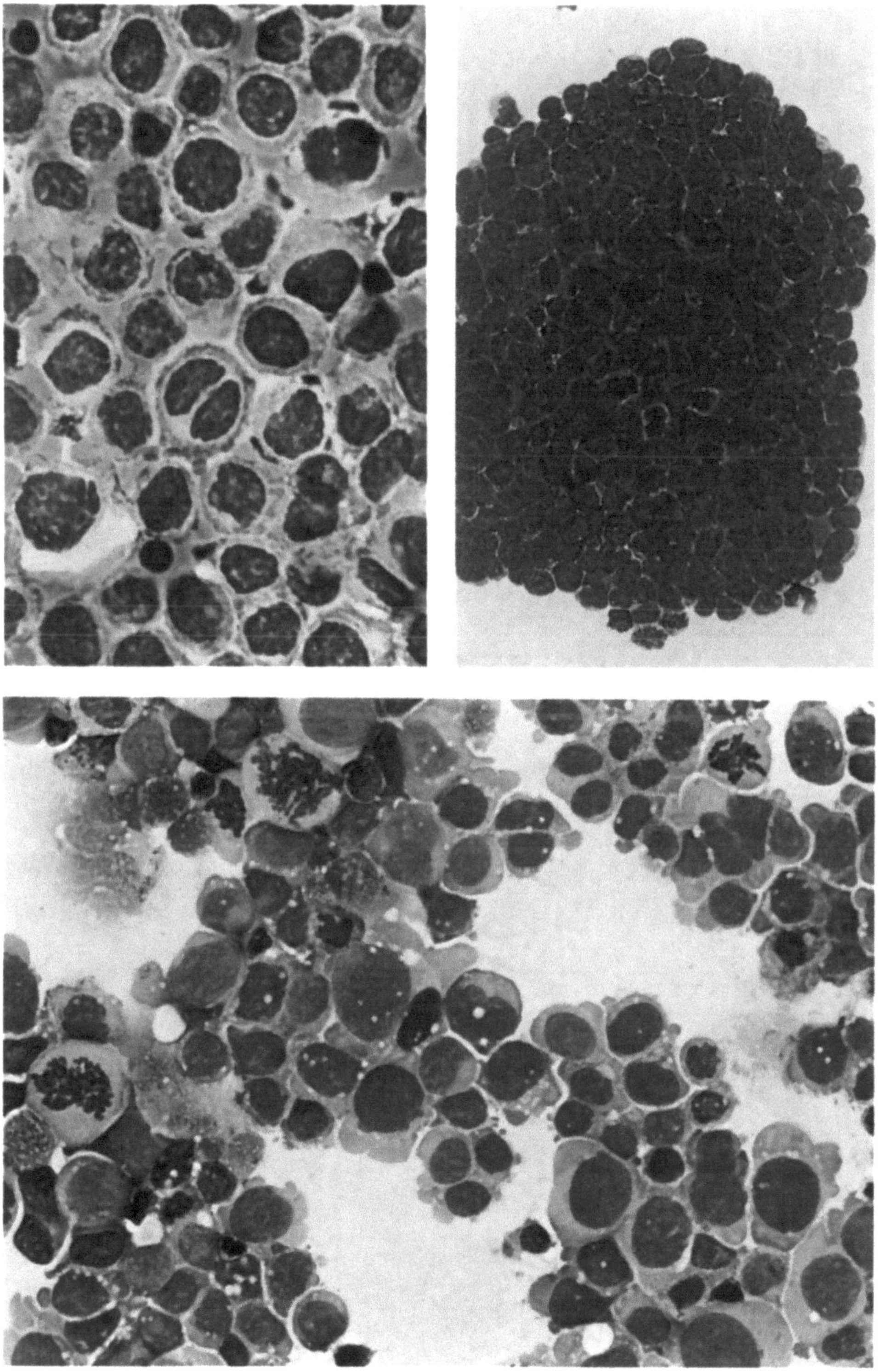

a c b

13.3. Leukemias

Cytologic monitoring of the CSF of leukemia patients has taken on extraordinary importance in recent years [1, 39, 99, 104, 128, 131, 196].

Findings which were familiar only to pathologists [110, 205] 10 or 20 years ago are no longer a clinical rarity, a result of constant improvement in chemotherapy, which has brought about a corresponding increase in the life expectancy of leukemia patients. Of special interest are findings of leukemic infiltrations of the meninges, which are genuine infiltrations (i.e., metastases) and not malignant neoplasia of existing cells [117].

These clinical findings can be explained by the special characteristics of the blood-CSF barrier, which is impermeable, or only slightly permeable, for almost all chemotherapeutic agents in current use. Standard medications can reach the blood and almost all organs, including the brain and the dura mater, in high cytotoxic concentrations. This does not apply to the CSF, and the result is a relatively uninhibited multiplication of leukemia cells which have either been shed from cell infiltrations of the leptomeninges or which have entered the CSF space directly from the blood. The leptomeninges are one of the most important breeding grounds for the leukemia cells which inundate the CSF and which initiate the general recurrence of the illness. It follows that the common expression "complete remission" can only apply when the CSF as well as the blood and bone marrow are free of leukemia cells.

Identification of leukemia cells in the CSF has diagnostic significance and a therapeutic consequence – intrathecal administration of appropriate chemotherapeutic agents and/or radiation therapy.

Tumor cells are sometimes found in the CSF in the initial stages of the disease, and this may be regarded as a prognostically unfavorable sign. Leukemia cells more often settle in the CSF after the illness has existed for a longer period, sometimes even after several remissions due to appropriate therapy and good response to the chosen chemotherapeutic agents. Meningitis leukemica often appears when an apparent remission has been achieved and the cytostatic dosage reduced. Our most recent experience indicates that meningitis leukemica appears in almost 40% of all stem cell leukemias. Though our results apply only to adults, the situation seems to be similar in children [16, 72, 131].

Cell counts in the CSF vary between normal values and more than $1{,}000/m^3$ without showing direct correlation with the severity of the illness. Higher values are rare. Cytologic analysis usually reveals a pure culture of cells from the specific leukemia; a meningitic cell reaction appears as a secondary finding, at best, and consists of monocytes or macrophages, small to middle-sized lymphocytes and a few neutrophils.

13.3.1. Acute Lymphoblastic Leukemia

Of the stem cell leukemias, the acute lymphoblastic form has the greatest tendency to metastasize to the CSF. This is primarily due to a good response to chemotherapy, which increases life expectancy and offers lymphoblasts an opportunity to settle in the CSF. The acute lymphoblastic form is the most common type of leukemia in children, but it is also frequently found in adults.

One can hardly find morphologic differences from other forms of leukemia when the May-Grünwald-Giemsa method is employed. The morphologic differences are never so discrete that classification could be attempted according to these criteria. Myeloid characteristics, such as Auer bodies or azurophilic granulation of the cytoplasm, are usually absent in acute lymphoblastic leukemia, and the cytoplasm more frequently contains vacuoles. The nuclei of the CSF cells are more polymorphic than those of the blood, bone marrow, and lymph nodes. It is sometimes impossible to differentiate from malignant lymphoma.

Cytochemical reaction to PAS stain is characteristic of acute lymphoblastic leukemia and often diagnostically decisive [53]. Lymphoblasts in the CSF show the same cytochemical behavior as those of blood and bone marrow. PAS-positive substances appear in the cytoplasm as a coarse-grained deposit which can resemble a string of pearls.

Plate 48

a–c Acute lymphoblastic leukemia. Patient: male, 14 years old. CSF: 330 cells/mm³.
(a, b) Pappenheim stain. (c) Acid phosphatase reaction.

d Acute lymphoblastic leukemia. Relapse of meningitis leukemica. Patient: female, 22 years old. CSF: 235 cells/mm³.
Lymphoblastic leukemia with conspicuously small cells.

e–g Acute lymphoblastic leukemia. Patient: female, 30 years old. CSF: 380 cells/mm³.
(e) Pappenheim stain. (f) PAS reaction.
(g) Signs of cell death after intrathecal chemotherapy (25 cells/mm³).

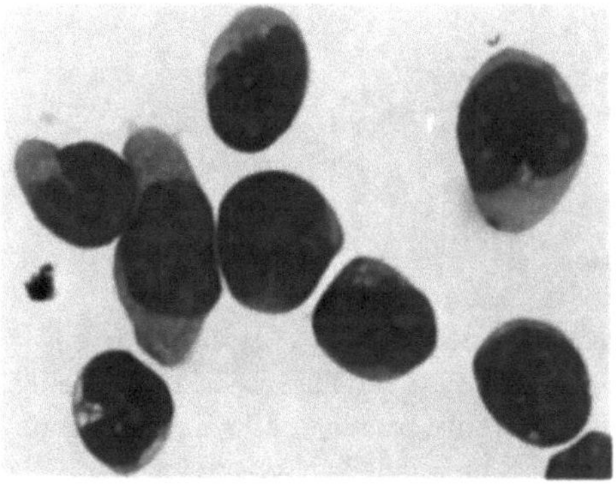
a

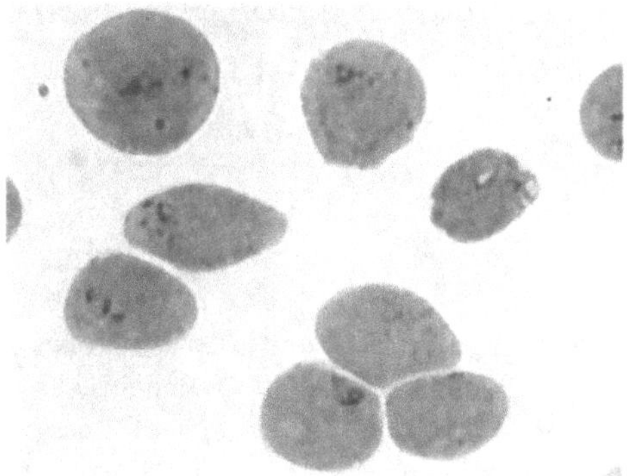
c

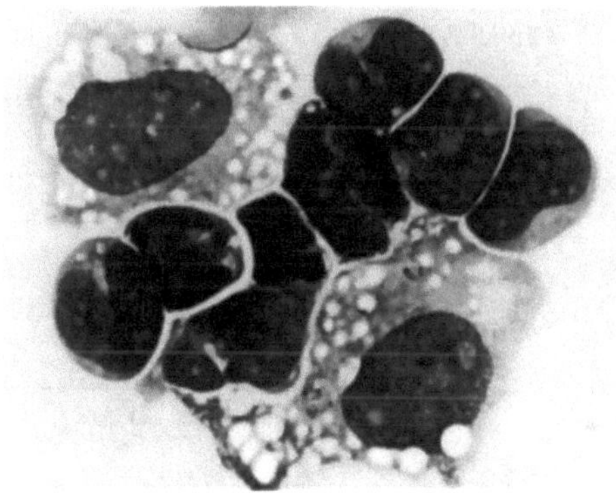
b

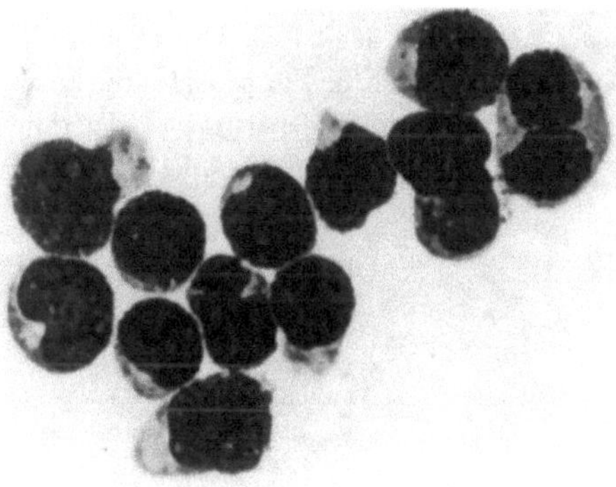
d

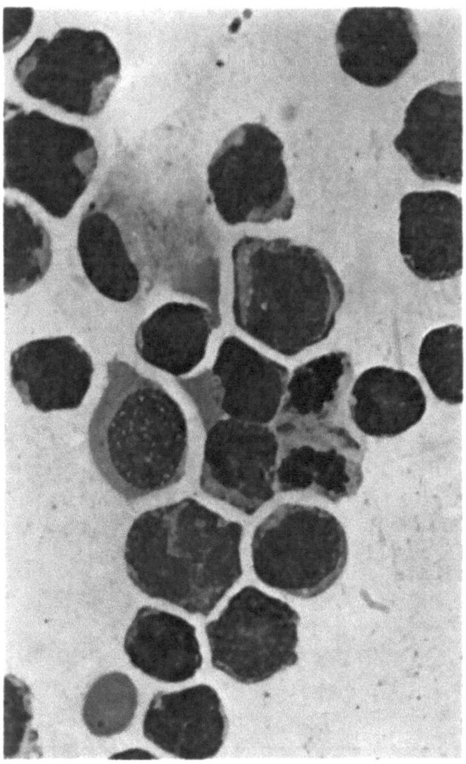
e

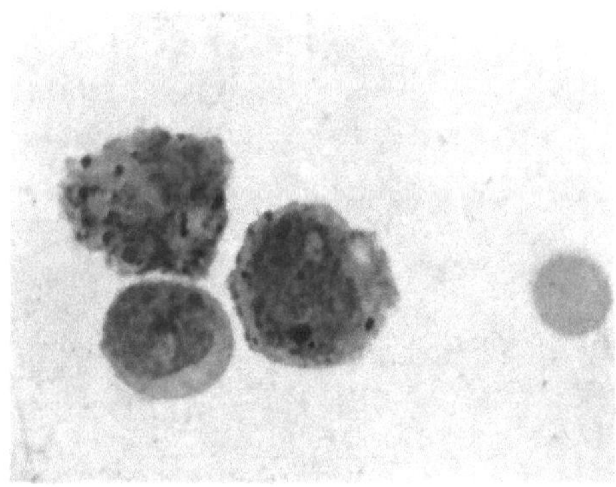
f

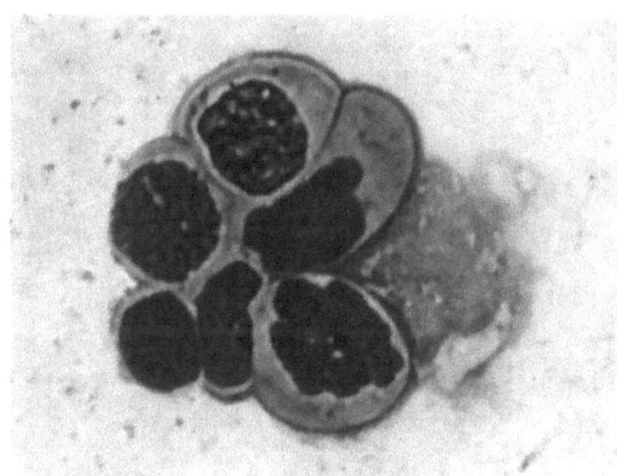
g

13.3.2. Acute Myeloblastic Leukemia

Myeloblastic leukemia occurs in all age groups but is more common in adults. Meningitis can sometimes accompany the first signs of illness. Panoptic staining reveals characteristics (azurophilic cytoplasmic granules, Auer bodies) which indicate myelogenic leukemia. However, only typical cytochemical behavior provides the basis for certain diagnosis. The peroxidase reaction is positive in the majority of myeloblastic leukemias, while PAS staining produces a slightly diffuse, positive reaction, at best. The esterase reaction is occasionally positive as well.

A number of stem cell leukemias cannot even be roughly classified according to morphologic and cytochemical criteria. These leukemia forms also often metastasize in the CSF.

Plate 49

a–c Myeloid leukemia. Patient: female, 45 years old. CSF: 140 cells/mm^3. (a) Pappenheim stain. (b) Peroxidase reaction. (c) Fluorescent stain with acridine orange.

d Stem cell leukemia. Cytochemical differentiation impossible. Patient: male, 44 years old. CSF: bloody, 220 cells/mm^3.

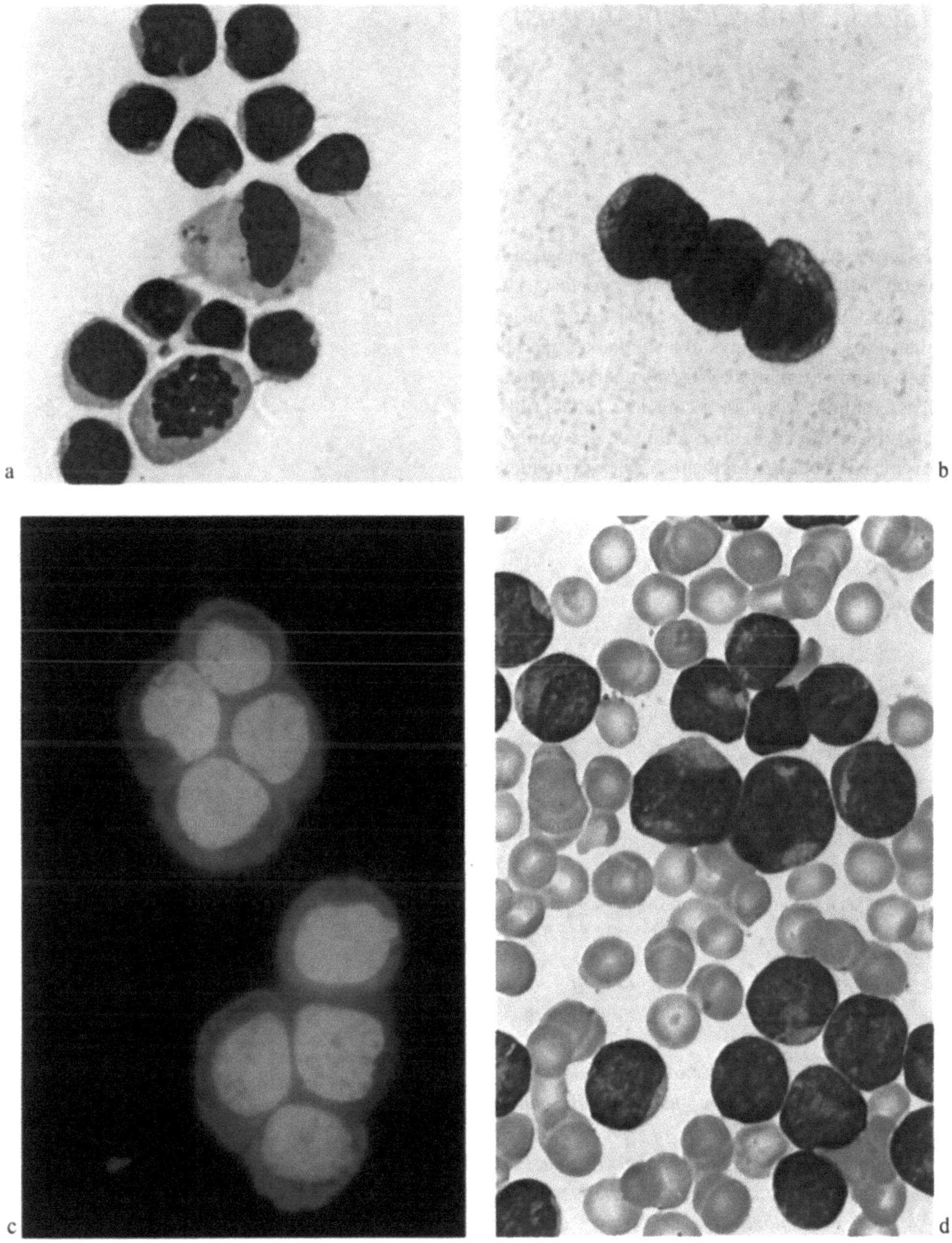
a
b
c
d

13.3.3. Promyelocytic Leukemia

Promyelocytic Leukemia. This prognostically unfavorable leukemia form often metastasizes to the CSF. The cells are consistently larger than those in acute myeloblastic leukemia. The cytoplasm is remarkable for its coarse, primarily basophilic granules, the nucleus for its polymorphism. Some mature granulocytes contain oversized nuclear segments. The peroxidase reaction is often positive.

Plate 50

a–c *Promyelocytic leukemia. Patient: male, 34 years old. CSF: 320 cells/mm³, 180 mg% protein.*
(a, b) Loose collection of tumor cells. Conspicuously large nucleoli. Basophilic stippling in cytoplasm of some relatively mature cells. Minimal phagocytic reaction. (c) Interference phase microscopy. × 1,250. Rare polyploid Nucleus. Fine granular texture of nuclei.

d, e *Promyelocytic leukemia. Patient: male, 36 years old. CSF: 20 cells/mm³. (c) Pappenheim stain. (d) Fluorescent stain with acridine orange × 800.*

f *Monocytic leukemia. Patient: male, 18 years old. CSF: 28 cells/mm³.*

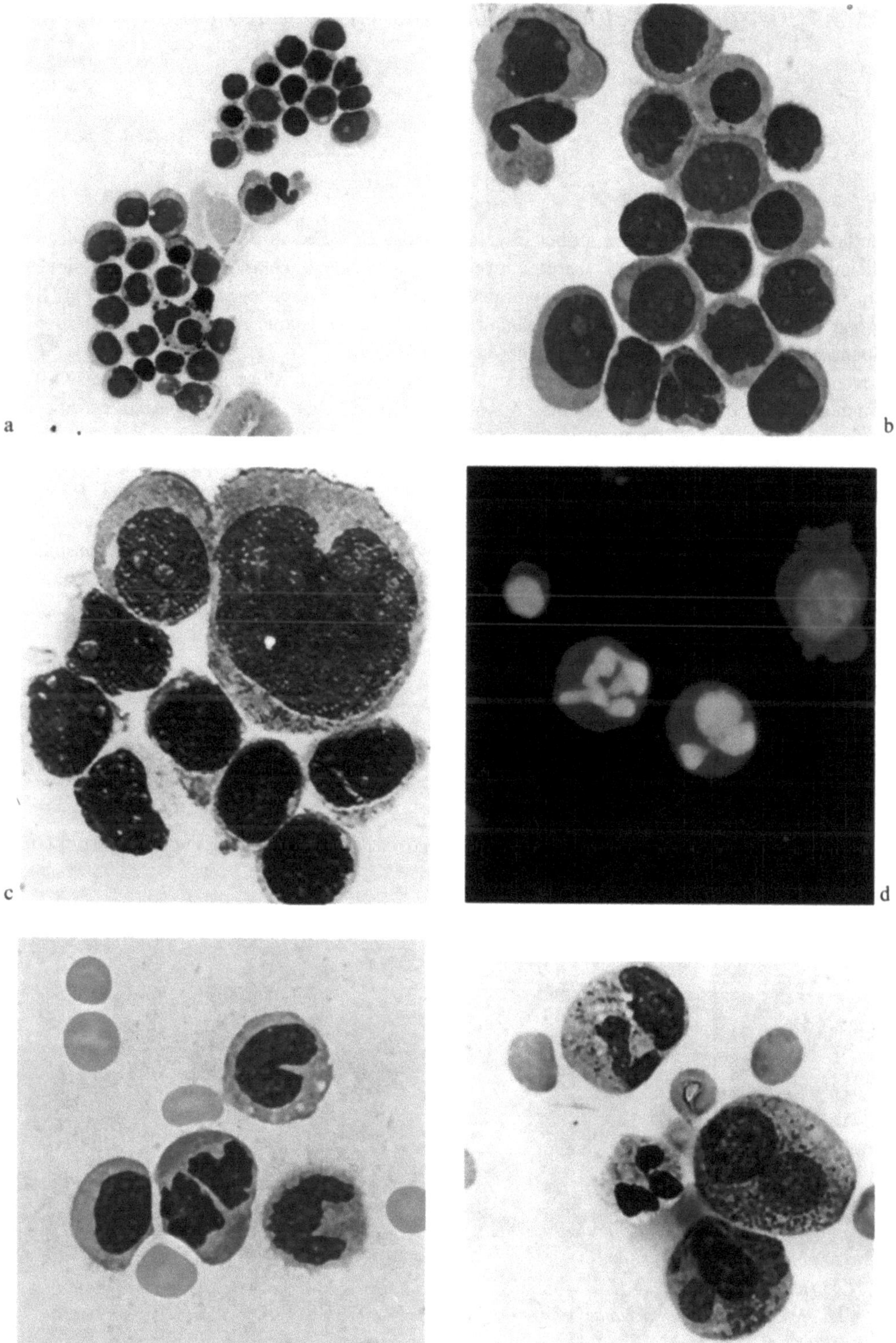
a
b
c
d
f
e

13.3.4. Chronic Myeloid and Chronic Lymphoid Leukemia

Both forms of leukemia occur principally in the middle-aged and the elderly. Chronic myeloid leukemia generally has a less favorable prognosis than chronic lymphoid leukemia. Spontaneous remissions can occur in the latter.

An increase in the number of basophils, eosinophils, promyelocytes and myeloblasts is conspicuous in chronic myeloid leukemia. Alkaline phosphatase activity in the leukocytes is reduced. Identification of the Philadelphia chromosome is decisive evidence for the presence of this form of leukemia. Meningitis is rare [107].

Meningitis is even less common in chronic lymphoid leukemia [35, 99]. This form of leukemia is also classified as a lymphoma of low degree of malignancy. The cellularity is very similar to that in lymphocytic meningitis, but the leukemia is characterized by nuclear polymorphism.

The cytoplasm stains rather weakly. Klein-Gumprecht shadow nuclei, typical of peripheral blood, are of little diagnostic use when found in the CSF, since they appear in a variety of illnesses. Individual cells may contain deposits of PAS-positive material in coarse granules. The lymphoblasts can proliferate in the terminal stage of the disease; this is probably the point at which the cells metastasize in the CSF.

Plate 51

Chronic lymphatic leukemia. Patient: female, 73 years old. CSF: 970 cells/mm³.

a, b Pappenheim stain.

c Fluorescent stain with acridine orange. ×800.

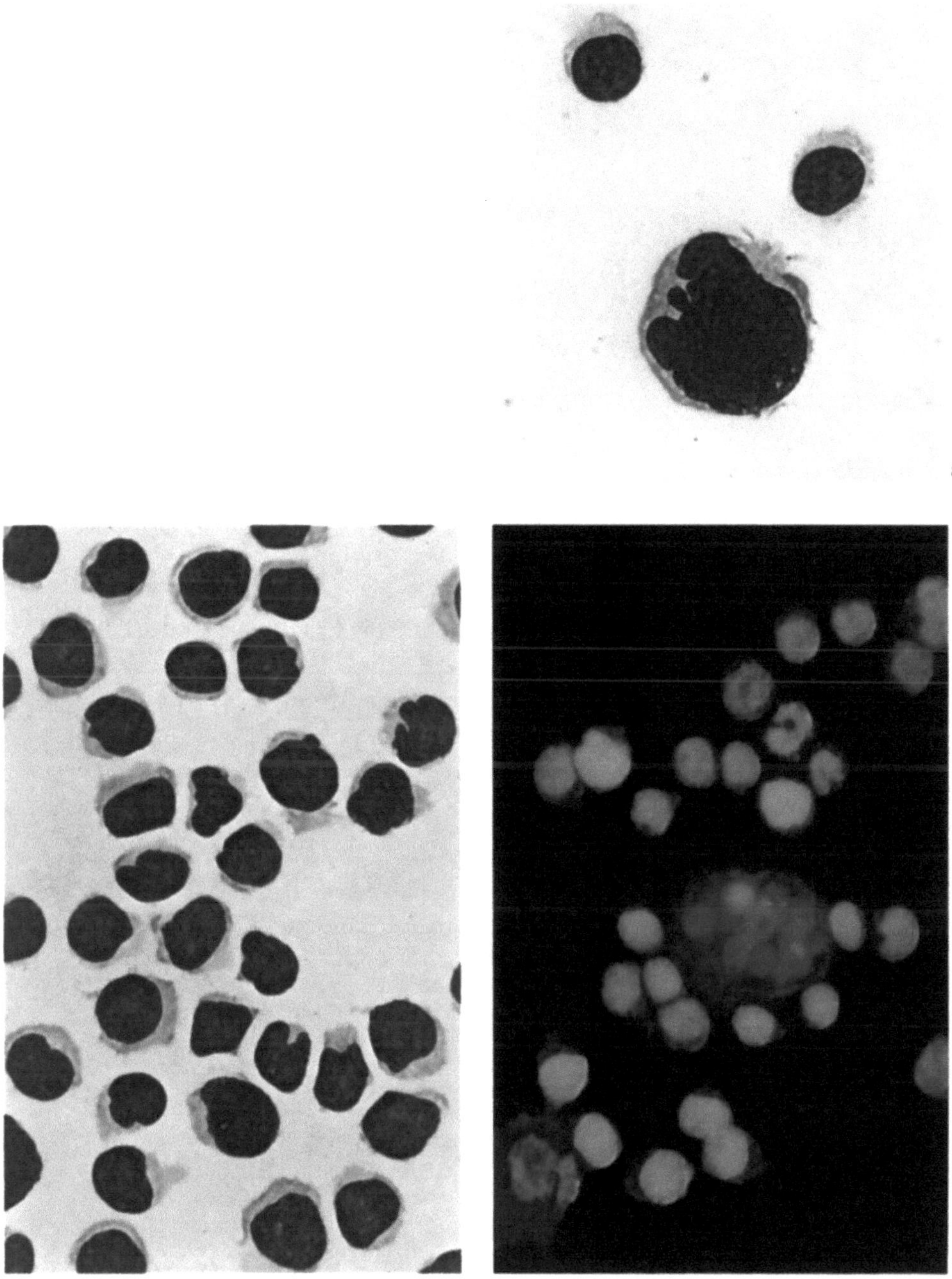

a

b

c

13.4. Malignant Lymphomas

Malignant lymphomas are rarer than leukemias, but they demonstrate the same tendency to infiltrate the meninges, so early diagnosis and special therapy have equal significance in CSF cytology [32, 63, 78, 91, 103, 128].

The category of malignant lymphomas includes Hodgkin's disease and all other malignant neoplasms of lympho-reticular tissue. Hodgkin tumors can be identified histologically with a high degree of certainty, but classification of other malignant lymphomas is more difficult. It may be that the classification proposed by LENNERT et al. [111] will prove most adequate to the variety of these tumors. In the latter system the terms reticulum cell sarcoma and lymphosarkoma—which are often applied to a number of different tumors—will be replaced by more exact terminology.

One can distinguish between lymphomas of greater and lesser degrees of malignancy. Chronic lymphocytic leukemia, immunocytoma (including macroglobulinemia) and centrocytoma (including Brill-Symmers disease) are examples of the less malignant type, while Burkitt's lymphoma, and lymphoblastic and immunoblastic sarcoma are considerably more malignant. All of the less malignant lymphomas have the capacity for becoming more malignant, and almost all malignant lymphomas can become leukemic and inundate the bone marrow and the blood. It is almost impossible to classify cells of malignant lymphomas in the CSF, in contrast to leukemias, which can usually be readly identified. The physician has to rely on clinical and histopathologic findings.

Positive CSF findings have been reported in Hodgkin sarcomas [103, 156]. We have found inflammatory meningeal reaction (up to 700 cells/mm^3) with some of these tumors, but we have never found Hodgkin cells or Reed-Sternberg cells. Burkitt's lymphoma appears to be limited to certain areas of the world and has an extraordinarily strong tendency to metastasize to the leptomeninges and the CSF [78, 237].

Lymphoblastic malignant lymphoma and immunoblastic malignant lymphoma (commonly called reticulum cell sarcoma) are most important in CSF cytology. Cytologic differentiation of these tumors in the CSF is rarely possible. The cells are quite uniform and most often appear singly, though rarely in cell clusters as well. Differential diagnosis is difficult in the latter case, especially differentiation from medulloblastoma. The cells almost always have a single nucleus, and the nuclear-cytoplasmic ratio is increased. Rare nuclear polymorphism has been observed in cells with round, horseshoe-shaped and segmented nuclei. 1 to 3 very prominent nucleoli are regularly present, and the mitotic index is high. Such CSF sometimes displays activation of phagocytes which have developed into macrophages of gigantic proportions. Their cytoplasm is packed with phagocytized tumor cells.

Some malignant lymphomas develop as primary brain tumors, while others appear in the process of metastatic invasion of the brain, the meninges or the CSF [79, 198]. One has to exclude the possibility of an extracranial tumor site when a primary intracranial lymphoma is suspected. The latter can develop in the brain itself or in the leptomeninges. Malignant cells always appear in the CSF in the case of leptomeningeal lymphoma, and a careful search usually leads to evidence of malignancy when cerebral growths are present. When lymphoma cells appear individually—not in clusters—it may be difficult to identify them among granulocytes, monocytes and macrophages. Sometimes, only the large nucleoli deviate from the physiological norm.

Plate 52

Meningitis in lymphoblastic sarcoma. Patient female, 6 years old. CSF: 7000 cells/mm^3.

Masses of immature cells. Irregular, often bean-shaped nuclei, with a maximum of 3 large nucleoli in most cases. Narrow border of cytoplasm. Cell in the middle of the picture is interspersed with fine vacuoles.

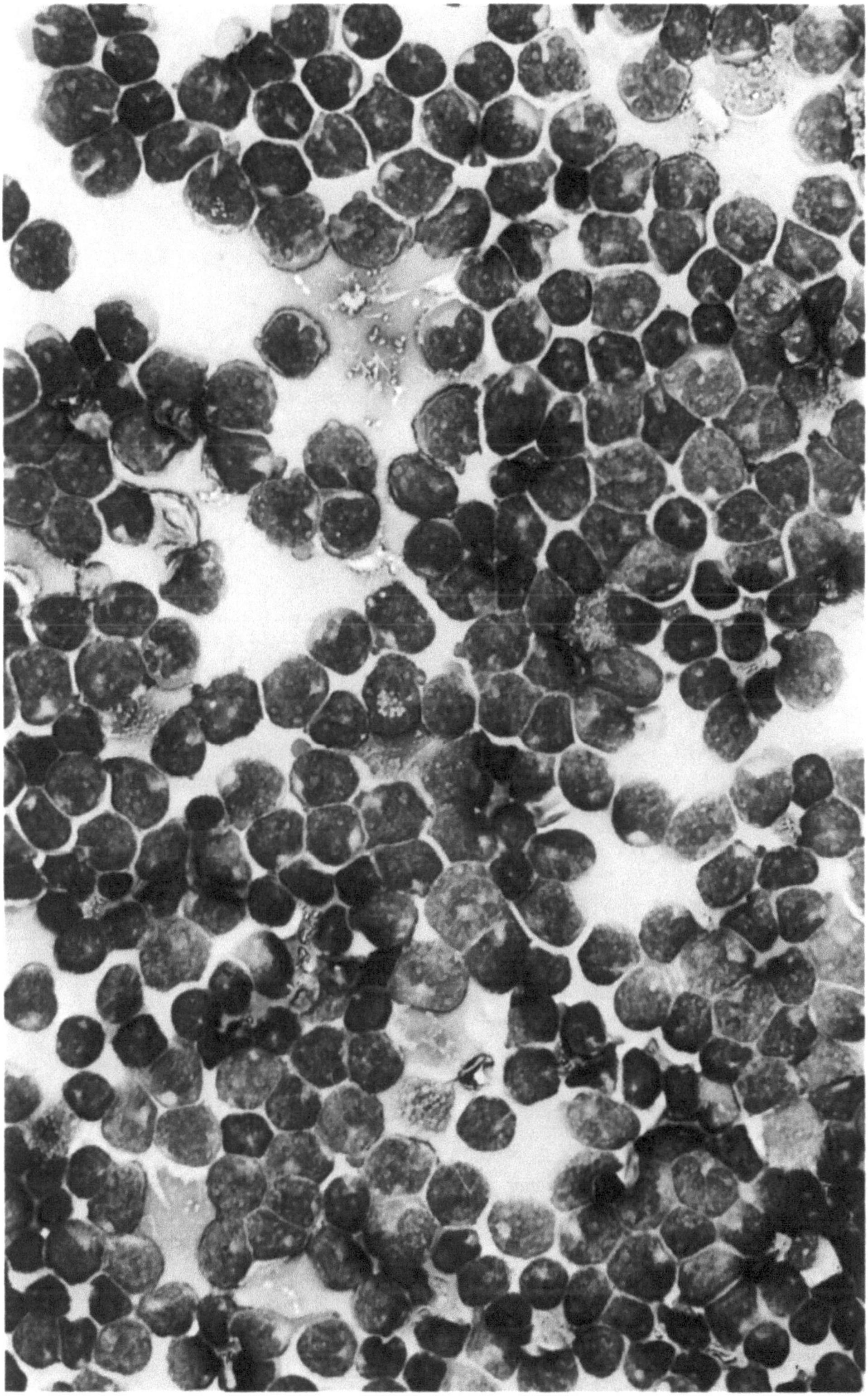

Plate 53

a *Same patient as Plate 51.*

b, c *Lymphoma similar to Burkitt's tumor. Patient: male, 18 years old. CSF: 210 cells/mm³, 130 mg% protein.*

d, e *Lymphoblastic sarcoma. Patient: male, 42 years old. CSF: 700 cells/mm³. (d) Pappenheim stain. (f) Fluorescent stain with acridine orange.*

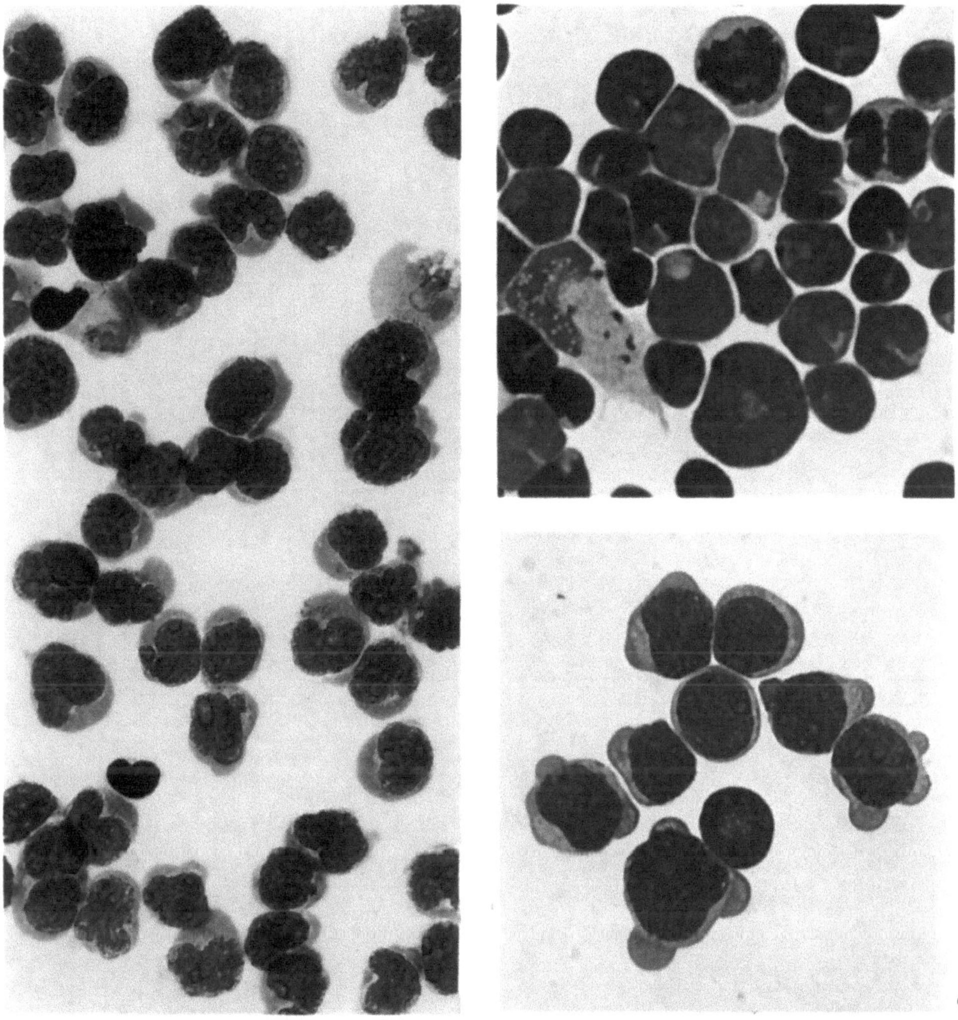

a

b

c

d

e

References

1. Aaronson, A.G., Hajdu, S.I., Melamed, M.R.: Spinal fluid cytology during chemotherapy of leukemia of the central nervous system in children. Am. J. clin. Path. **63**, 528–537 (1975).
2. Afifi, A.M.: Myeloma cells in the cerebrospinal fluid in plasma cell neoplasia. J. Neurol. Neurosurg. Psychiat. **37**, 1162–1165 (1974).
3. Aksel, I.S.: Au sujet d'une forme confusionelle-encephalitique de la maladie de Behçet. Etude anatomo-clinique et électro-éncéphalographique Rev. neurol. **99**, II 316–322 (1958).
4. Alema, G., Bignami, A.: Involvement of the nervous system in Behçet's disease. In: Int. Symp. on. Behçet's Disease, Rome. Basle: Karger, 1966, pp. 52–56.
5. Alzheimer, A.: Einige Methoden zur Fixierung der zelligen Elemente der Cerebrospinalflüssigkeit. Zbl. Nervenhk. **30**, 449–451 (1907).
6. Aronson, S.M., Garcia, J.H., Aronson, B.E.: Metastatic neoplasms of the brain: Their frequency in relation to age. Cancer **17**, 558–563 (1964).
7. Aronson, A.S., Garwicz, S., Sörnäs, R.: Cytology of the cerebrospinal fluid in children with acute lymphoblastic leukemia. J. Ped. **85**, 222–224 (1974).
8. Ashton, P.R., Hollingsworth, A.S., Johnston, W.N.: The cytopathology of metastatic breast cancer. Acta cytol. **19**, 1–6 (1976).
9. Babinski, J., Nageotte: Contribution à l'étude du cytodiagnostic du liquide céphalo-rachidien dans les affections nerveuses. Bull. Soc. méd. Hôp. Paris **18**, 537–545 (1901).
10. Balzereit, F.: Liquorzytologie in der neurologischen Tumordiagnostik. Internist **7**, 122–127 (1966).
11. Bammer, H.: Zur Tumorzelldiagnostik im Liquor cerebrospinalis. Dtsch. Z. Nervenhk. **185**, 89–109 (1963/64).
12. Bammer, H.: Über die Beziehung der Plasmazellen im Liquor zur Aktivität der multiplen Sklerose. Verh. Dtsch. Ges. inn. Med. **32**, 733–736 (1966).
13. Bannwarth, A.: Die Zellen der Cerebrospinalflüssigkeit. Arch. Psych. Nervenkr. **100**, 533–573 (1933).
14. Baringer, J.R.: A simplified procedure for spinal fluid cytology. Arch. Neurol. (Chicago) **22**, 305–308 (1970).
15. Behçet, H.: Über rezidivierende, aphthöse, durch einen Virus verursachte Geschwüre am Mund, am Auge und an den Genitalen. Dermat. Wschr. **105**, 1152–1157 (1937).
16. Bennet, J.M., Ruberg, J., Dixon, S.: A new semiautomated method for the concentration of cerebrospinal fluid for cytologic examination. Amer. J. Clin. Path. **50**, 533–536 (1968).
17. Bernhardt, H., Gourley, R.D., Young, J.M., Shepherd, M.C., Kilian, J.J.: A modified membrane-filter technique for detection of cancer cells in body fluids. Amer. J. Clin. Path. **36**, 462–464 (1961).
18. Bertalanffy, L. von, Masin, M., Masin, F.: A new and rapid method for diagnosis of vaginal and cervical cancer by fluorescence microscopy. Cancer **11**, 873–887 (1958).
19. Bischoff, A.: Der derzeitige Stand der Liquor-Cytodiagnostik. Schweiz. Med. Wschr. **90**, 479–487 (1960).
20. Bischoff, A.: Erfahrungen mit der Tumorzelldiagnostik im Liquor cerebrospinalis. Acta Neurochir. **9**, 510–524 (1961).
21. Bischoff, A.: Das Vorkommen von Plasmazellen im Liquor cerebrospinalis bei der Multiplen Sklerose. Dtsch. Z. Nervenhk. **185**, 606–617 (1964).
22. Bots, G.T.A.M., Went, L.N., Schaberg, A.: Results of a sedimentation technique for cytology of cerebrospinal fluid. Acta Cytol. **8**, 234–241 (1964).
23. Braunsteiner, H.: Funktion der Leukozyten. Internist (Berl.) **3**, 89–95 (1962).
24. Brucher, J.M., Smidts, M., Lecuit, J.: L'analyse cytologique due liquide céphalorachidien par une méthode de cytocentrifugation. Résultats dans la sclerose en plaques. Acta neurol. belg. **72**, 201 (1972).
25. Burgmann, G.P.: Die zytologische Diagnostik der Hirngeschwülste. In: Symposion über die Zerebrospinalflüssigkeit, Sayk, J. (ed.). Jena: VEB Gustav Fischer, 1966, pp. 35–38.
26. Burton, J.F.: Tumor cells in cerebrospinal fluid. Med. Radiogr. Photogr. **37**, 22–23 (1961).
27. Cardozo, P.L.: Advantages and disadvantages of the May-Grünwald Giemsa procedure in exfoliative cytology. Acta Cytol. (Philad.) **2**, 284 (1958).
28. Casperson, T.O.: Cell Growth and Cell Function. New York: Norton, 1950.

29. CHESTER, D.C., EMERY, J.L., PENNY, S.R.: Fat-laden macrophages in cerebrospinal fluid as an indication of brain damage in children. J. Clin. Path. **24**, 753–756 (1971).
30. COLLOMB, A., MORIN, G.: Gliome cérébral avec réction pléocytaire importante du liquide céphalo-rachidien. Presse méd. (1956) 1841–1842.
31. CHRISTOPHERSON, N.M.: Prognosis of breast cancer based on pathologic type. Cancer **24**, 1179–1181 (1969).
32. CURRIE, S., HENSON, R.A.: Neurological syndromes in the reticuloses Brain. **94**, 307–320 (1971).
33. DANNER, E.D., GMELICH, J.T.: A comparative study of tumor cells from metastatic carcinoma of the breast in effusions. Acta cytol. **19**, 509–518 (1975).
34. DELALLA, E.R.: NBT-Test in bacterial meningitis. Lancet **1**, 747–748 (1972).
35. DEN HARTOG JAGER, W.A.: Cytopathology of the cerebrospinal fluid examined with the sedimentation technique after Sayk. J. Neurol. Sci. **9**, 155–177 (1969).
36. DOMMASCH, D.: Monocytes and histiocytes in cell cultures of cerebrospinal fluid. Morphology of cultured CSF cells. J. Neurol. **209**, 103–114 (1975).
37. DOMMASCH, D., PRZUNTEK, H., GRÜNINGER, W., MERTENS, H.G.: Intrathecal cytostatic chemotherapy of meningitis carcinomatosa. Clinical manifestation and cerebrospinal fluid cytology in a case of metastatic carcinoma of the breast. Eur. Neurol. **14**, 178–191 (1976).
38. DOS REIS, J.B., MOTA, I., BEI, A., DOS REIS FILHO, J.B., ZUKERMAN, E.: Os basófilos do liquido cefalorraqueano. Arq. neuropsiquiat. (San Paulo) **31**, 10 (1973).
39. DREWINKO, B., JULLIVAN, M.P., MARTIN, T.: Use of the cytocentrifuge in the diagnosis of meningeal leukemia. Cancer **31**, 1331–1336 (1973).
40. DROPMANN, K.: Beitrag zur Liquorzelldiagnostik. Dtsch. Z. Nervenhk. **178**, 131–140 (1958).
41. DUFOUR, M.H.: Meningite sarcomateuse diffuse avec envahissement de la moelle et des vaccines, cytologie positive et spéciale du liquide céphalorachidien. Rev. Neurol. **12**, 104–106 (1904).
42. DUFRESNE, J.J.: Praktische Zytologie des Liquors. Ciba-Geigy AG, Basel 1973
43. DUVE, C. DE: Lysosomes, a new group of cytoplasmic particles. In: Subcellular Particles. New York: Ronald Press, 1959.
44. DYKEN, P., MULINA, J.: Acridine orange fluorescence of cells of the cerebrospinal fluid. Clin. Res. **22**(1): 93, 1974.
45. ECKES, K.H., MUTSCHLER, D.: Veränderungen des Liquorzellgehaltes während fraktionierter Encephalographie. Nervenarzt **24**, 302–303 (1953).
46. EL-BATATA, M.: Cytology of cerebrospinal fluid in the diagnosis of malignancy. J. Neurosurg. **28**, 317–326 (1968).
47. ENESTRÖM, S.: Suction technique for sedimentation of body fluid cells. Acta Path. Microbiol. Scand. **59**, 473–480 (1963).
48. ENESTRÖM, S.: Meningeal relation of primary brain tumours. Cancer **19**, 1117–1123 (1966).
49. ENGELHARDT, P., PATZOLD, U.: Liquorzytologischer Nachweis von primären und sekundären Geschwülsten des Zentralnervensystems. J. Neurol. **209**, 45–57 (1973).
50. ENGELHARDT, P.: Diagnostischer Wert der Siderophagen im Liquorcytogramm. J. Neurol. **208**, 201–206 (1975).
51. ENGELHARDT, P.: Intrathecal cytostatic therapy of meningeal carcinomatosis. Autoradiographic investigations of the CSF cells. J. Neurol. **213**, 309–315 (1976).
52. FARR, G.H.: Exfoliative cytology of metastatic neuroblastoma. Acta Cytol. **76**, 203–206 (1972).
53. FELDGES, A.J.: Diagnosis of meningeal leukemia in acute childhood lymphocytic leukemia with periodic acid Schiff reaction of cytocentrifuged liquor. Acta haemat. (Basel) **49**, 154–158 (1973).
54. FIKRIG, S.: NBT dye test and differential diagnosis of meningitis. J. Pediatr. **82**, 855–857 (1973).
55. FISCHER, O.: Klinische und anatomische Beiträge zur Frage nach den Ursachen und der Bedeutung der zerebrospinalen Pleocytose (der Zellvermehrung im Liquor cerebrospinalis). Jb. Psych. Neurol. **27**, 313–359 (1906).
56. FISCHER, S.: Diagnostic cytology by an extended membrane filter technique. Acta Path. Microbiol. Scand. Suppl. **212**, 33–37 (1970).
57. FISCHER-WILLIAMS, M., BOSANQUET, F.D., DANIEL, P.M.: Carcinomatosis of the meninges. A report of 3 cases. Brain **78**, 42–58 (1955).
58. FORSTER, E.: Die Bedeutung des Liquorzellbildes für die-Diagnostik der Tumoren des Zentralnervensystems um die vom Plexus und den Meningen ausgehenden Tumoren. Z. ges. Neurol. Psych. **126**, 683–720 (1930).
59. FUJITA, S., KITAMURA, T.: Origin of Brain Macrophages and the Nature of the So-Called Microglia. Acta Neuropath. (Berlin) Suppl. VI, 291–296 (1975).
60. FUKUI, M., YAMAKAWA, Y., YAMASAKI, T., KITAMURA, K., TABIRA, T., SADOSHIMA, S.: ^{3}H-thymidine autoradiography of CSF cells in primary reticulum cell sarcoma of the brain. J. Neurol. **210**, 143–150 (1975).
61. GILL, G.W.: Comparative Filter Technique Acta Cytologica. Vol. **19**, 3, 207–209 (1975).
62. GRABLE, E.: Nuclear size of cells in normal stomachs in gastric atrophy and in gastric cancer. Gastroenterol. **32**, 1104–1112 (1957).

63. GRIFFIN, J.W., THOMPSON, R.W., MITCHINSON, M.J., DE KIEWIET, J.C., WELLAND, F.H.: Lymphomatous leptomeningitis. Am. J. Med. **51**, 200–208 (1971).
64. GREGER, J., WIECZOREK, V.: Über das Vorkommen von Plasmazellen im Liquor cerebrospinalis bei neurologischen Erkrankungen. Wien. Z. Nervenhk. **23**, 366–374 (1966).
65. GUILLAIN, G., VERDUN, M.: La forme meningée des tumeurs cérébrales. Bull. Soc. méd. Hôp. (Paris) **32**, 521–534 (1911).
66. GUSEO, A.: Über die Makrophagen des Liquor cerebrospinalis. Z. Neurol. **200**, 136–147 (1971).
67. GUSEO, A.: Morphological Signs as Indications of Function of Cells in the Cerebrospinal Fluid. J. Neurol. **212**, 159–170 (1976).
68. HACKENBERG, P.: Kritische Erwägungen zur Frage der „maskierten" Makrophagen im Liquorzellsediment. Dtsch. Z. Nervenhk. **192**, 124–138 (1967).
69. HACKENBERG, P., KOSS, L.G.: Kritische Bemerkungen zur Bedeutung der Lymphoidzelle im Liquor der akuten Meningitis. Dtsch. Z. Nervenhk. **190**, 284–294 (1967).
70. HAJDU, S.I.: Cytologic diagnosis of metastatic myosarcomas. Acta cytol. (Balt.) **13**, 545–551 (1969).
71. HANSEN, H.H.: The cyto-centrifuge and cerebrospinal fluid cytology. Acta cytol. (Balt.) **18**, 259–262 (1974).
72. HARMS, D.: Der diagnostische Wert der Liquorzytologie. Klin. Pädiat. **187**, 142–150 (1975).
73. HERNDON, R.M.: The electron microscopic study of cerebrospinal fluid sediment. J. Neuropath. exp. Neurol. **28**, 128 (1969).
74. HOFFMANN, G.T., EARLE, K.M.: Meningioma with malignant transformation in the subarachnoid space. J. Neurosurg. **17**, 486–492 (1960).
75. HRAZDIRA, C.L.: Riesenzellige Makrophagen mit einer möglichen regulativen Tätigkeit im Liquor cerebrospinalis. Dtsch. Z. Nervenheilk. **194**, 37–50 (1968).
76. IIVANAINEN, M., TASKINEN, E.: Differential cellular increase in cerebrospinal fluid after encephalography in mentally retarded patients. J. Neurol. Neurosurg. Psychiat. **37**, 1252–1258 (1974).
77. ITO, U., INABA, Y.: Electron microscopic observation of cerebro-spinal fluid cells—A new method for embedding of CSF cells. Electronmicroscopy **19**, 265–270 (1970).
78. JANOTA, I.: Malignant Lymphoma-Cells in the Cerebrospinal fluid. Lancet **1964 II**, 677–678.
79. Jellinger, K., Wrchovszky, A.: Liquorzytologische Befunde bei malignen Lymphomen des Zentralnervensystems. Wien. klin. Wschr. **85**, 522–525 (1973).
80. JOHNSTON, W.W., GINN, F.L., AMATULLI, J.M.: Light and electron microscopic observation of malignant cells in cerebrospinal fluid from metastatic alveolar cell carcinoma. Acta cytol. **15**, 365–371 (1971).
81. JUNKER, F.: Die Zellen des Liquor cerebrospinalis im Phasenkontrastmikroskop. Dtsch. Z. Nervenhk. **166**, 237–246 (1951).
82. KAFKA, V.: Über Technik und Bedeutung der zytologischen Untersuchung des Liquor cerebrospinalis. Mschr. Psych. Neurol. **27**, 414–438 (1910).
83. KAFKA, V.: Über die Polynucleose im Liquor cerebrospinalis bei der progressiven Paralyse. Z. ges. Neurol. Psych. **1**, 648–659 (1910).
84. KALM, H.: Über die Metastasierung von Geschwülsten in die Liquorräume. Dtsch. Z. Nervenhk. **159**, 397–407 (1948).
85. KALM, H.: Ein malignes Tentorium Meningiom mit Metastasierung in die Medulla oblongata und in die subarachnoidalen Liquorräume. Dtsch. Z. Nervenhk. **163**, 131–140 (1950).
86. KALM, H.: Phasenkontrastbeobachtungen an Hirngeschwülsten. Dtsch. Z. Nervenhk. **179**, 267–184 (1959).
87. KAPLOW, L.S.: Cytochemical Heterogeneity of Human Circulating Monocytes. Acta cytol. **19**, 358–365 (1975).
88. KELLER, H.U., SORKIN, E.: Chemotaxis of leukocytes. Experientia (Basel), **42**, 641–652 (1968).
89. KENDEL, K., MEIER-EWERT, K.: Die Diagnose von Hirnmetastasen aus dem Liquorzellausstrich. Dtsch. med. Wschr. **93**, 2075–2076 (1968).
90. KERNOHAN, J.W.: Tumors of the central nervous system. In: Atlas of Tumor Pathology. Washington, DC: AFIP, 1952.
91. KING, D.K., LOH, K.K., AYALA, A.G., GAMBLE, J.F.: Eosinophilic meningitis and lymphomatous meningitis. Ann. intern. Med. **82**, 228 (1975).
92. KISTLER, G.S.: Zur Membranfilter-Technik in der Cytodiagnostik des Liquor cerebrospinalis. Nervenarzt **41**, 507–510 (1970).
93. KISTLER, G.S.: Zur exfoliativen Cytologie kleiner Flüssigkeitsmengen. Schweiz. med. Wschr. **92**, 863–866 (1972).
94. KNIERIM, G.: Über diffuse Meningealcarcinose mit Amourose und Taubheit bei Magenkrebs. Beitr. path. Anat. **44**, 409–429 (1908).
95. KLINE, T.S.: Cytological examination of the cerebrospinal fluid. Cancer **15**, 591–597 (1962).
96. KLINE, T.S.: Tumor cells in the cerebrospinal fluid. J. Neurosurg. **19**, 679–684 (1962).
97. KLINGE, O.: Riesenzellenenzephalitis bei Candida-Infektion. Nervenarzt **35**, 356–360 (1964).
98. KÖLMEL, H.W., CHONÉ, B.: Tumorzellen im Liquor—Ein Beitrag zur differenzierten Cytodiagnostik. Nervenarzt **43**, 644–645 (1972).
99. KÖLMEL, H.W., PAAL, G., DRINGS, P.: Meningosis leucaemica des Erwachsenen. Nervenarzt **44**, 527–536 (1973).

100. Kölmel, H.W.: A method for concentrating cerebrospinal fluid cells. Acta cytol. **21**, 154–157 (1977).
101. Kolar, O., Zeman, W.: Spinal fluid cytomorphology. Description of apparatus, technique and findings. Arch. Neurol. (Chicago) **18**, 44–51 (1968).
102. Kolář, O., Zeman, W., Ciembroniewicz, J., Vesely, F.: Über die Bedeutung der eosinophilen Leukozyten bei neurologischen Krankheiten. Wien. Z. Nervenheilk. **27**, 97–106 (1969).
103. Kolář, O.J.: Differential diagnostic aspects in malignant lymphomas involving the central nervous system. Acta neuropath. (Berlin) Suppl. VI: 181–186 (1975).
104. Komp, D.M., Cox, B.J.: Cytocentrifugation in the management of central nervous system leukemia. J. Pediat. **81**, 992–994 (1972).
105. Krentz, M.J., Dyken, P.R.: Cerebrospinal fluid cytomorphology. Sedimentation vs Filtration. Arch. Neurol. **26**, 253–257 (1972).
106. Kubie, L.S., Schultz, G.M.: Vital and supravital studies of the cells of the cerebrospinal fluid and of the meninges in cats. Bull. Johns Hopk. Hosp. **37**, 91–129 (1925).
107. Kwaan, H.C., Pierre, R.V., Long, D.L.: Meningeal involvement as first manifestation of acute myeloblastic transformation in chronic granulocytic leukemia. Blood **33**, 348–352 (1969).
108. Langevoot, H.L., Cohn, Z.A., Hirsch, J.G., Humphrey, J.H., Spector, W.G., Furth, R. van: The nomenclature of mononuclear phagocytic cells. Proposal of a new classification. R. van Furth, Ed., Mononuclear phagocytes. Blackwell, Oxford-Edinburgh (1970).
109. Larson, C.P.: Cytologic diagnosis of tumor cells in cerebrospinal fluid. J. Neurosurg. **10**, 337–341 (1953).
110. Leidler, F., Russel, W.O.: The brain in leukemia. Arch. Path. **40**, 14–33 (1945).
111. Lennert, K.: Morphology and Classification of Malignant Lymphomas and So-called Reticuloses. Acta Neuropath. (Berlin) Suppl. VI, 1–16 (1975).
112. Lesse, S., Netsky, M.G.: Metastasis of neoplasms to the central nervous system and meninges. Arch. Neurol. Psych. **72**, 133–153 (1954).
113. Ludwin, S.K., Conley, F.K.: Malignant meningioma metastasing through the cerebrospinal pathways. J. Neurol. Neurosurg. Psychiat. **38**, 136–142 (1975).
114. Macias Sanchez, R., Ordo Nez, S.: Cysticercosis cerebralis diagnostico clinico, radiologica y de laboratorio, prognostico. Analisis de 186 casos. Prensa med. mex **35**, 6–14 (1970).
115. Marks, V., Marrack, D.: Tumor cells in the cerebrospinal fluid. J. Neurosurg. Psych. **23**, 194–201 (1960).
116. Marks, V., Marrack, D.: Technique for staining of cells from C.S.F. Confinia Neurol. **20**, 310–315 (1960).
117. Mastrangelo, R., Zuelzer, W., Ecklund, P.S., Thompson, R.T.: Chromosomes in the spinal fluid: evidence for metastatic origin of meningeal leukemia. Blood **35**, 227–235 (1970).
118. Mazalton, A., Le Beau, J.: Intérêt de l'étude cytologique du liquide cephalo-rachidien dans les affections malignes du système nerveux central. Neurochir. **13**, 219–227 (1970).
119. Mazalton, A., Daum, S.: Cytologie du liquide céphalo-rachidien. Neurochir. **17**, Suppl. 1, 89–109 (1971).
120. McCall, C.R., Caves, J., Cooper, R., DeChatelet, L.: Functional Characteristics of Human Toxic Neutrophils. The Journ. Of Infectious Diseases **124**, 68–75 (1971).
121. Mc Cormack, L.J., Hazard, J.B., Gardner, W.J., Klotz, J.G.: Cerebrospinal fluid changes in secondary carcinoma of the meninges. Amer. J. Clin. Path. **23**, 470–478 (1953).
122. Mc Cormack, L.J., Hazard, J.B., Belovich, D., Gardner, W.J.: Identification of neoplastic cells in cerebrospinal fluid by a wet-film method. Cancer **10**, 1293–1299 (1957).
123. Mc Cormick, W.F.: A membrane filter technique for cytology of spinal fluid. Amer. J. Clin. Path. **38**, 191–197 (1962).
124. Mc Menemey, W.H., Cumings, J.N.: The value of the examination of the cerebrospinal fluid in the diagnosis of intracranial tumours. J. Clin. Path. **12**, 400–411 (1959).
125. Meissner, G.F.: Carcinoma of stomach with meningeal carcinosis: report of 4 cases. Cancer **6**, 313–318 (1953).
126. Möbius, W.: Zystizerkose des Gehirns und Eosinophile im Liquor. Dtsch. Med. Wschr. **24**, 1318–1323 (1970).
127. Mollaret, M.P.: La méningite endothelio leucocytaire multirecurrente au maladie nouvelle (Documents cliniques). Rev. Neurol. **76**, 57–76 (1944).
128. Mouriquand, C., Augusseau, S., Bachelot, C., Breyton, M.: Interet de l'examen cytologique du liquide cephalo rachidien dans la surveillance des leucemies aigues et de lymphosarcomes. Lyon Med. **231**, 315 (1974).
129. Naylor, B.: An exfoliative cytologic study of intracranial fluids. Neurol. **11**, 560–570 (1961).
130. Naylor, B., Path, M.C.: The cytologic diagnosis of cerebrospinal fluid. Acta Cytol. **8**, 141–149 (1964).
131. Nies, B.A., Malmgren, R.A., Chu, E.W., Del Vecchio, P.R., Thomas, L.B., Freireich, E.J.: Cerebrospinal fluid cytology in patients with acute leukemia. Cancer (Philad.) **18**, 1385–1391 (1965).
132. Nomarski, G.: Microinterférometre differentielle a ondes polarisées. J. Phys. Radium **16**, 9 (1955).

133. NONNE, M.: Syphilis und Nervensystem. 3rd ed. Berlin: Karger 1915.
134. OBRADOR, S.: Cysticercosis cerebri. Acta Neurochir. (Vienna) **10**, 320–364 (1962).
135. OEHMICHEN, M., SCHÜTZE, G.: Erythrophagen in der Liquorzelldiagnostik der Subarachnoidalblutung. Nervenarzt **44**, 407–416 (1973).
136. OEHMICHEN, M., GRÜNINGER, H., GENCIC, M.: Experimental studies on kinetics and functions of mononuclear phagocytes of the central nervous system. Acta neuropath. (Berlin), Suppl. VI: 285–290 (1975).
137. OEHMICHEN, M.: Cerebrospinal fluid cytology. Stuttgart: G. Thieme 1976.
138. OLISCHER, R.M.: Die Zellreaktionen des Liquor cerebrospinalis. Ergebnisse liquorzytologischer Untersuchungen in der Klinik entzündlicher und Tumorerkrankungen des Zentralnervensystems. Habilitationsschrift, Rostock (1969).
139. OLISCHER, R.M.: Liquorcytochemische Methoden und Ergebnisse. In: Symposion über die Zerebrospinalflüssigkeit SAYK, J. (ed.). Jena: VEB G. Fischer 1966, pp. 79–88.
140. OLISCHER, R.M., SAYK, J.: Liquor-, insbesondere Zellbefunde bei Meningitiden. Fortschr. Neurol. Psych. **35**, 453–474 (1967).
141. OLISCHER, R.M.: Der Nachweis der unspezifischen Esterase in Liquorzellen. Z. Neurol. **200**, 61–69 (1971).
142. OLISCHER, R.M.: Die PAS-Reaktion in der Cytodiagnostik des Liquor cerebrospinalis. Psych. Neurol. Med. Psych. **24**, 716–725 (1972).
143. OLSON, M.E., CHERNIK, N.L., POSNER, J.B.: Infiltration of the leptomeninges by systemic cancer. Arch. Neurol. (Chic.) **30**, 122–137 (1974).
144. OPPENHEIM, H.: Über Hirnsymptome bei Karzinomatose ohne nachweisbare Veränderung im Gehirn. Charité-Ann. **13**, 334–344 (1888).
145. PAPANICOLAOU, G.N.: Atlas of Exfoliative Cytology. Cambridge, Mass.: Harvard Univ. Press 1954.
146. PARAG NIKOLIC, M., ARKO, K., PUSKARIC, N.H.: The applic. of Prussian blue reaction for the proof of siderophages in cerebrospinal fluid (Jugosl.). Acta med. Jugosl. **28**, 3 (1974).
147. PARK, B.M.: Infection and nitroblue-tetrazolium reduction by neutrophils. Lancet **II**, 532–534 (1968).
148. PÉTER, A.: The plasma cells of the cerebrospinal fluid. J. neurol. Sci. **4**, 227–239 (1967).
149. PENZHOLZ, H.: Die metastatischen Erkrankungen des Zentralnervensystems bei bösartigen Tumoren. Acta Neurochir. Suppl. XVI. Springer, 1–205 (1968).
150. PETTE, H.: Über diffuse Karzinose der weichen Hirnrückenmarkshäute. Dtsch. Z. Nervenhk. **74**, 226–233 (1922).
151. PEIFFER, J., SCHWARZ, E.J.: Beitrag zur Enzymhistochemie der Liquor- und Leptomeningealzellen. Nervenarzt **42**, 267–269 (1971).
152. PLATT, W.R.: Exfoliative-cell diagnosis of central nervous system lesions. Arch. Neurol. Psych. (Chicago) **66**, 119–144 (1951).
153. QUADRA, M.: Hirncysticercose und Eosinophilie des Liquor cerebrospinalis. Rev. Neuropsichiatr. **12**, 339 (1949).
154. QUINCKE, H.: Die Lumbalpunktion des Hydrocephalus. Klin. Wschr. **28**, 929–932 (1891).
155. RAUTENBACH, M.: Liquorzytologie 1. Die Zellarten des Liquor cerebrospinalis. Internist prax. **10**, 75–81 (1970).
156. RAWLINSON, D.-G., BILLINGHAM, M.E., BERRY, P.F., KEMPSON, R.L.: Cytology of the Cerebrospinal Fuid in Patients with Hodgkin's Disease or Malignant Lymphoma. Acta Neuropath. (Berlin) Suppl., VI, 187–191 (1975).
157. REHM, O.: Atlas der Zerebrospinalflüssigkeit. Jena: Fischer 1932.
158. REYNAUD, A.J.: A new filter for diagnostic cytology. Acta Cytol. **11**, 289–294 (1967).
159. RICH, J.R.: A survey of cerebrospinal fluid cytology. Bull. Los Angeles Neurol. Soc. **34**, 115–131 (1969).
160. RICH, J.R.: A membrane filter technique for cerebrospinal fluid cytology. J. Neurosurg. **36**, 661–666 (1972).
161. RINDFLEISCH, W.: Über diffuse Sarkomation der weichen Hirn- und Rückenmarkshäute mit charakteristischen Veränderungen der Cerebrospinalflüssigkeit. Dtsch. Z. Nervenhk. **26**, 135–148 (1904).
162. SAYK, J.: Ergebnisse neuer liquorcytologischer Untersuchungen mit dem Sedimentierkammerverfahren. Ärzt. Wschr. **9**, 1042–1046 (1954).
163. SAYK, J.: Klinischer Beitrag zur Liquor-Eosinophilie und Frage der allergischen Reaktion im Liquorraum. Dtsch. Z. Nervenhk. **177**, 62–72 (1957).
164. SAYK, J., WIECZOREK, V.: Über besondere Verlaufsformen der Meningitis. Zur Frage der retothelialen Riesenzellmeningitis. Arch. Psych. Nervenkr. **200**, 182–196 (1960).
165. SAYK, J.: Cytologie der Cerebrospinal Flüssigkeit. Jena: Fischer Verlag 1962.
166. SAYK, J.: The cerebrospinal fluid in brain tumours. In: Handbook of Clinical Neurology. VINKEN, P.J., BRUYN, G.W. (eds.). Amsterdam: North-Holland Publ. 1974, Vol. XIII.
167. SAYK, J., OLISCHER, R.M.: Fortschritte der Liquorzytologie bei der Diagnostik bösartiger Hirngeschwülste. III. Mitteilung. Psychiat. Neurol. med. Psychol. (Lpz) **19**, 88–99 (1967).
168. SCARLATO, G., SMIRNE, S.: Moderne vedute in tema di citologia liquorale. Riv. Neurol. **37**, 258–288 (1967).

169. SCHALTENBRAND, G.: Die chronischen Meningitiden. Dtsch. Z. Nervenhk. **171**, 275–297 (1954).
170. SCHILLING, V.: Das Blutbild und seine klinische Verwertung. Jena: Gustav Fischer, 1943.
171. SCHLOTE, W., ROOS, W.: Gibt es ein charakteristisches Liquorzellbild bei der Multiplen Sklerose? Nervenarzt **45**, 576–587 (1974).
172. SCHMIDT, R.M., SCHMIDT, TH., SEIFERT, B.: Beitrag zur Fluoreszenz- und Electronenmikroskopie der Liquorzellen. Psychiatr. Neurol. med. Psychol. (Lpz.) **18**, 201–203 (1966).
173. SCHMIDT, R.M.: Der Liquor cerebrospinalis während der Behandlung der Neurolues. Dtsch. med. J. **16**, 190–196 (1965).
174. SCHMIDT, R.M.: Zur Bedeutung der Lymphoidzellen im Liquor cerebrospinalis bei entzündlichen Erkrankungen des Zentralnervensystems. Schweiz. Arch. Neurol. Neurochir. Psychiat. **113**, 295–301 (1973).
175. SCHMIDT, R.M.: Der Liquor cerebrospinalis VEB Volk und Gesundheit. Berlin 1968.
176. SCHÖNENBERG, H.: Das Liquorzellbild bei der Meningitis tuberkulosa. Mschr. Kinderhk., **97**, 509–515 (1949).
177. SCHÖNENBERG, H.: Eine einfache Methode zur Herstellung gut differenzierbarer Liquorzellpräparate. Dtsch. Med. Wschr. **74**, 881 (1949).
178. SCHÖNENBERG, H.: Über das Liquorzellbild bei der idiopathischen mononucleären Meningitis. Z. Kinderhk. **69**, 274–284 (1951).
179. SCHÖNENBERG, H.: Untersuchungen über Vitalspeicherung in Liquorzellen sowie Phasenkontrast- und Fluoreszenzmikroskopischen Beobachtungen über den Zerfall der Liquorzellen. Z. Kinderhk. **72**, 157–180 (1952).
180. SCHOLZ, H., SUMMER, K.: Klinisch-diagnostischer Wert liquorcytologischer Befunde bei Entzündungen der Meningen. Wien. Z. Nervenhk. **28**, 283–305 (1970).
181. SCHÜMMELFEDER, N.: Histochemical significance of the polychromatic fluorescence induced in tissues stained with acridine orange. J. Histochem. Cytochem. **6**, 392 (1958).
182. SCHWARZE, E.W.: Zur Differenzierung der im Liquor cerebrospinalis des Menschen anzutreffenden Zellelemente. Inaug. Diss. Tübingen 1967.
183. SCHWARZE, E.W.: Zytomorphologischer Vergleich der „kleinen und großen Rundzellen" des Liquor cerebrospinalis mit Lymphozyten des peripheren Blutes. Verhdlg. d. dt. Gesellsch. f. Pathologie **57**, 143–148 (1973).
184. SEAL, S.H.: A method for concentrating cancer cells suspended in large quantities of fluid. Cancer **9**, 866–868 (1956).
185. SEAL, S.H.: A sieve for the isolation of cancer cells and other large cells from the blood. Cancer **17**, 637–642 (1964).
186. SIMON, G., SCHRÖER, H.: Ein neues Verfahren zur vollständigen Erfassung der im Liquor vorkommenden Zellen (Zellfangverfahren). Arch. Psych. Nervenhk. **204**, 74–85 (1963).
187. SIMON, G., SCHRÖER, H.: The cell-catch procedure. J. Neurosurg. **20**, 787–792 (1965).
188. SÖRNÄS, R.: Transformation of mononuclear cells in cerebrospinal fluid. Acta Cytol. **15**, 545–552 (1971).
189. SÖRNÄS, R.: The cytology of the normal cerebrospinal fluid. Acta Neurol. Scand. **48**, 313–320 (1972).
190. SÖRNÄS, R., ÖSTLUND, H., MÜLLER, R.: Cerebrospinal fluid cytology after stroke. Arch. Neurol. (Chicago) **26**, 489–501 (1972).
191. SPAAR, F.-W., BAUER, H.J.: Fluoreszenzmikroskopischer Nachweis von Tumorzellen im Liquor cerebrospinalis mit Akridinorange. Zeiss-Information **19**, 19–22 (1971).
192. SPIELMEYER, W.: Histopathologie des Nervensystems. Berlin: Springer, 1922, Vol. I, p. 199.
193. SPRIGGS, A.J.: Malignant cells in cerebrospinal fluid. J. Clin. Path. **7**, 122–130 (1954).
194. SPRIGGS, A.J.: Myeloid cells in cerebrospinal fluid. J. Neurol. Neurosurg. Psych. **21**, 305–307 (1958).
195. SPRIGGS, A.I., BODDINGTON, M.M.: The Cytology of Effusions in the Pleural, Pericardial and Peritoneal Cavities and of Cerebrospinal Fluid. 2nd ed. London: William Heinemann 1968.
196. SPRIGGS, A.J., BODDINGTON, M.M.: Leukaemic cells in cerebrospinal fluid. Brit. J. Haemat. **5**, 83–91 (1959).
197. STAMMLER, A., MARGUTH, F., SCHMIDT-WITTKAMP, E.: Die Meningitis carcinomatosa und sarcomatosa. Fortschr. Neurol. Psychiat. **32**, 53–77 (1964).
198. STEFANKO, S., MOFFO, D.: Primary reticulum-cell-sarcoma of the brain. A clinicopathological study. Clin. Neurol. Neurosurg. **77**, 96–106 (1974).
199. STÖRTEBECKER, T.P.: Metastatic tumors of the brain from a neurosurgical point of view. A follow-up study of 158 cases. J. Neurosurg. **11**, 84–111 (1954).
200. STROUTH, J.C., DYKEN, M.: Encephalopathy of Behçet's disease. Neurol. (Minneap.) **14**, 794–805 (1964).
201. SUBERO, A., FONCIN, J.F., LEBEAU, J.: Diagnostic cytologique du liquide céphalorachidien par la chambre de sédimentation de Suta. Neurochir. (Paris) **14**, 627–634 (1968).
202. SVIEN, H.J., GATES, E.M., KERNOHAN, J.W.: Spinal subarachnoid implantation associated with ependymoma. Arch. Neurol. Psych. (Chicago) **62**, 847–856 (1949).
203. TARLOV, J.M., DAVIDOFF, L.: Subarachnoid and ventricular implants in ependymal and other gliomas. J. Neuropath. **5**, 213–224 (1946).
204. THALMANN, H.: Zellveränderungen im Liquor

cerebrospinalis nach Pneumoencephalogramm und Myelogramm. Nervenarzt **39**, 129–131 (1968).

205. THOMAS, L.B.: Pathology of leukemia in the brain and meninges: post mortem studies of patients with acute leukemia and of mice given inoculations of L 1210 leukemia. Cancer Res. **25**, 1555–1571 (1965).
206. TOMPKINS, V.N., HAYMAKER, W., CAMPELL, E.H.: Metastatic pineal tumors. J. Neurosurg. **7**, 159–169 (1950).
207. TORVIK, AN.: The Relationship between Microglia and Brain Macrophages. Experimental Investigations. Acta Neuropath. (Berlin) Suppl. VI, 297–300 (1975).
208. TOURTELOTTE, W., QUAN, K.C., HAERER, A., BRYAN, E.R.: Neoplastic cells in the cerebrospinal fluid. Report of a case of metastatic melanoblastoma. Neurology **13**, 866–868 (1963).
209. TOURTELOTTE, W.W.: Multiple sclerosis cerebrospinal fluid. In: Handbook of Clinical Neurology. Vinken, P.J., Bruyn, G.W. (eds.), Amsterdam: North-Holland Publ. 1970, Vol. XIII.
210. TRELLES, J.O.: Cerebral cysticercosis. Wld. Neurol. **2**, 488–497 (1961).
211. ULBRICHT, H.: Phasenkontrastmikroskopische Liquorzelluntersuchungen II. Zerfallene Rundzellen in der Hirn-Rückenmarksflüssigkeit bei Syphilis. Die Hypothese von Rundzellen und die Ursachen für ihren Zerfall. Arch. Dermat. Syph., **194**, 308–313 (1952).
212. VAHAR-MATIAR, H., MÜLLER, J., LIPPES, G.: Der blutige Liquor. Acta Neurochir. **29**, 229–245 (1973).
213. VAN ECK, J.H., GO, K.G., EBELS, E.J.: Metastatic tumours of the brain. Psychiat. Neurol. Neurochir. (Amst.) **68**, 443–462 (1965).
214. VOSS, G.: Über die Liquorcytologie des frühesten Stadiums der Multiplen Sklerose. Nervenarzt **12**, 470–471 (1939).
215. WAGNER, J.A., FROST, J.K., WISOTZKEY, H., JR.: Subarachnoid neoplasia: incidence and problems of diagnosis. South. Med. J. **53**, 1503–1508 (1960).
216. WALT, F.: A medulloblastoma in an infant: abnormal cells in the cerebrospinal fluid. Arch. Dis. Child. **14**, 84–86 (1939).
217. WATSON, P.: A slide centrifuge: An apparatus for concentrating cells in suspension onto a microscopic slide. J. Lab. Clin. Med. **68**, 491–501 (1966).
218. WENDER, M., SNIATALA, M.: Zur Feinstruktur der Liquorzellen. Wien. Z. Nervenhk. **27**, 38–44 (1969).
219. WERNER, T.: Ein Pinealom mit diffuser Metastasierung in die Meningen. Zbl. Neurochir. **4**, 155–160 (1939).
220. WESEMANN, W.: Methodischer Beitrag zur Liquorzelldiagnostik. Dtsch. Z. Nervenhk. **191**, 360–366 (1967).
221. WERTLAKE, P.T., MARKOVITS, B.A., STELLAR, S.: Cytologic evaluation of cerebrospinal fluid with clinical and histologic correlation. Acta Cytol. (Balt.) **16**, 224–239 (1972).
222. WIDAL, F., SICARD, J.A., RAVAUT, P.: Cytodiagnostic de la méningite tuberculeuse. C.R. hebd. Soc. Biol. **52**, 838–843 (1900).
223. WIECZOREK, V.: Liquorveränderungen bei Blutungen in den Subarachnoidalraum mit besonderer Berücksichtigung des Liquorzellbildes. Dtsch. Z. Nervenhk. **186**, 87–100 (1964).
224. WIECZOREK, V.: Liquorzytologie der Geschwülste. In: Symposion über die Zerebrospinalflüssigkeit, J. SAYK (ed.) Jena: VEB Gustav Fischer, 1966, pp. 39–51.
225. WIECZOREK, V., STAHL, J., BOCK, R.: Zum Vorkommen von Riesenzellen im Liquor cerebrospinalis. Dtsch. Z. Nervenhk. **912**, 246–264 (1968).
226. WIECZOREK, V., HEID, R., BOCK, R.: Ungewöhnliche Liquorzellveränderungen bei einem Kleinhirntumor. Nervenarzt **42**, 270–273 (1971).
227. WIECZOREK, V.: Erfahrungen mit der Tumorzelldiagnostik im Liquor cerebrospinalis bei primären und metastatischen Hirngeschwülsten. Dtsch. Z. Nervenhk. **186**, 410–432 (1964).
228. WIECZOREK, V.: Zum Vorkommen von Lymphoidzellen im Liquor cerebrospinalis. Psych. Neurol. **154**, 384–397 (1967).
229. WILKINS, R.H., ODOM, G.L.: Ependymal-choroidal cells in cerebrospinal fluid. Increased incidence in hydrocephalic infants. J. Neurosurg. **41**, 555–560 (1974).
230. WILKINS, R.H., ODOM, G.L.: Cytological changes in cerebrospinal fluid associated with resections of intracranial neoplasms. J. Neurosurg. **25**, 24–36 (1966).
231. WOLF, C.: Des éléments de diagnostic tirés de la ponction lombaire. Diss. Paris, 1901.
232. WOLF, S.M., SCHOTTLAND, D.L., PHILLIPS, L.L.: Involvement of nervous system in Behçet's syndrome. Arch. Neurol. **12**, 315–325 (1965).
233. WOLFF, H.: Eosinophile Meningitis. Nervenarzt **27**, 225–226 (1956).
234. WOODRUFF, K.H.: Cerebrospinal fluid cytomorphology using cytocentrifugation. Amer. J. Clin. Path. **60**, 621–627 (1973).
235. YAMAKAWA, Y., FUKUI, M., OHTA, H., KITAMURA, K.: H3-thymidine autoradiography of the CSF cells in cases of non-neoplastic disease. J. Neurol. **211**, 195–202 (1976).
236. ZERNICKE, F.: Das Phasenkontrastverfahren bei der mikroskopischen Beobachtung. Phys. Z. **36**, 848–851 (1935).
237. ZIEGLER, J.L., BLUMING, A.Z., FASS, L., MORROW, R.H., IVERSEN, O.H.: Burkitt's Lym-

phoma Cell Kinetics Treatment and Immunology. VIII. Human Leukemia and Lymphoma. Unifying Concepts of Leukemia, Bibl. haemat. No. 3, pp. 1046–1052. Basel: Karger 1973.

238. ZÜLCH, K.J.: Die Hirngeschwülste in biologischer und morphologischer Darstellung. Leipzig: J.A. Barth, 1956.

Subject Index

Related Titles

K.J. Zülch, **Atlas of Gross Neurosurgical Pathology**
379 figures. V, 228 pages. 1975. ISBN 3-540-06480-X
Distribution rights for Japan: Nankodo Co. Ltd., Tokyo

V. Chan-Palay, **Cerebellar Dentate Nucleus** (Organization, Cytology and Transmitters)
293 figures, incl. 79 plates, some in color. XXI, 548 pages. 1977. ISBN 3-540-07958-0

S.L. Palay / V. Chan-Palay, **Cerebellar Cortex** (Cytology and Organization)
267 figures, incl. 203 plates. X, 348 pages. 1974. ISBN 3-540-06228-9
Distribution rights for Japan: Igaku Shoin Ltd., Tokyo

Cerebral Localization (An Otfrid Foerster Symposium)
Editors: K.J. Zülch, O.D. Creutzfeldt, G.C. Galbraith. 95 figures. X, 339 pages. 1975
ISBN 3-540-07379-5

Central Nervous System (Studies on Metabolic Regulation and Function)
Editors: E. Genazzani, H. Herken. Scientific Committee: E. Beccari, G. Cavallo,
J.-P. Changeux, G. Filogamo, E. Genazzani, H. Herken, P. Mandel
Local Organizing Secretary: R. Di Carlo
121 figures, 55 tables. VIII, 249 pages. 1974. ISBN 3-540-06444-3

Brain Edema (Pathophysiology and Therapy)
Cerebello Pontine Angle Tumors (Diagnosis and Surgery)
Editors: K. Schürmann, M. Brock, H.-J. Reulen, D. Voth. 187 figures.
XVII, 385 pages. 1973. ISBN 3-540-06486-9.
Distribution rights for Japan: Nankodo Co. Ltd., Tokyo
(Advances in Neurosurgery, Volume 1)

Cerebral Angiomas (Advances in Diagnosis and Therapy)
Editors: H.W. Pia, J.R.W. Gleave, E. Grote, J. Zierski. 161 figures. X, 285 pages. 1975
ISBN 3-540-07073-7. Distribution rights for Japan: Maruzen Co., Tokyo

Cerebral Blood Flow (Clinical and Experimental Results)
International Symposium on the Clinical Applications of Isotope Clearance Measurement of Cerebral Blood Flow, Mainz, April 1969.
Editors: M. Brock, C. Fieschi, D.H. Ingvar, N.A. Lassen, K. Schürmann.
113 figures, XX, 291 pages. 1969. ISBN 3-540-04436-1

Cerebral Circulation and Metabolism (6th International CBF Symposium, June 6–9, 1973)
Editors: T.W. Langfitt, L.C. McHenry jr., M. Reivich, H. Wollman.
180 figures, 100 tables. XXVIII, 566 pages. 1975. ISBN 3-540-06645-4
Distribution rights for Japan: Nankodo Co., Tokyo

Springer-Verlag Berlin Heidelberg New York

FSC
www.fsc.org
MIX
Papier aus verantwortungsvollen Quellen
Paper from responsible sources
FSC® C105338

If you have any concerns about our products,
you can contact us on
ProductSafety@springernature.com

In case Publisher is established outside the EU,
the EU authorized representative is:
Springer Nature Customer Service Center GmbH
Europaplatz 3, 69115 Heidelberg, Germany

Printed by Libri Plureos GmbH
in Hamburg, Germany